WRITING HUMAN FACTORS
RESEARCH PAPERS

Writing Human Factors Research Papers

A Guidebook

DON HARRIS
HFI Solutions Ltd, UK

ASHGATE

Published by
Ashgate Publishing Limited
Wey Court East
Union Road
Farnham
Surrey, GU9 7PT
England

Ashgate Publishing Company
110 Cherry Street
Suite 3-1
Burlington
VT 05401-3818
USA

www.ashgate.com

British Library Cataloguing in Publication Data
Harris, Don, 1961–
   Writing human factors research papers : a guidebook.
   1. Technical writing. 2. Human engineering--Authorship.
   I. Title
   808'.06662-dc23

   ISBN: 978-1-4094-4000-0 (hbk)
   ISBN: 978-1-4094-3999-8 (pbk)
   ISBN: 978-1-4094-4001-7 (ebk)

Library of Congress Cataloging-in-Publication Data
Harris, Don, 1961–
   Writing human factors research papers : a guidebook / by Don Harris.
      p. cm.
   Includes bibliographical references and index.
   ISBN 978-1-4094-4000-0 (hardback : alk. paper) -- ISBN 978-1-4094-3999-8 (pbk. : alk. paper) -- ISBN 978-1-4094-4001-7 (ebook) 1. Technical writing--Psychological aspects. I. Title.
   T11.H337 2011
   808.066--dc23

                                                                          2011035965

Reprinted 2015

Printed in the United Kingdom by Henry Ling Limited, at the Dorset Press, Dorchester, DT1 1HD

Contents

List of Figures

List of Tables

Acknowledgements

I have tried to illustrate most of the points that I make in this book with examples from papers that I have written with many different colleagues over the years. I hope that this brings the issues that I am talking about to life. I am extremely grateful to these people because without them none of this would have been possible. In particular, I have to say that I have learned more from Neville Stanton than from anyone else when it comes to understanding what makes a good Journal paper. Some of the knowledge was even acquired while we weren't in a bar.

I have also learned a great deal from being close to the publication process itself, so I also need to thank all the journal and book publishers that I have worked with, but particularly Guy Loft and all the people from Ashgate.

The papers contained in Appendices 1 and 2:

Harris, D. & Maxwell, E. (2001). Some considerations for the development of effective countermeasures to aircrew use of alcohol while flying. *International Journal of Aviation Psychology*, *11*, 237–252.

and

Harris, D., Chan-Pensley, J. & McGarry, S. (2005). The development of a multidimensional scale to evaluate motor vehicle dynamic qualities. *Ergonomics*, *48*, 964–982.

are reprinted by kind permission of Taylor & Francis Ltd (www.tandf.co.uk/journals).

Finally, I have to thank Fiona for all her help and support over the years. I really couldn't have done it without her. And of course, I must also mention Megan who has also managed to tolerate me (so far)!

About the Author

Don Harris PhD BSc is Director of HFI Solutions ltd. He is also a Visiting Professor at Shanghai Jiao Tong University (China) and an Honorary Visiting Fellow at the University of Leicester. Don is a Fellow of the Institute for Human Factors and Ergonomics, a Chartered Psychologist and a Registered European Aviation Psychologist.

Don has an extensive publication record. He has published more than 70 scientific journal papers; 100 conference papers and has edited or written 16 text books on Human Factors in Aviation and Defence. His latest book is *Human Performance on the Flight Deck,* recently published by Ashgate.

Over the last 25 years Don has been Editor-in-Chief of the journal *Human Factors and Aerospace Safety* and is currently Editor-in-Chief of the new scientific journal *Aviation Psychology and Applied Human Factors*. He has also acted as Guest Editor for special issues of *Ergonomics*; the *International Journal of Cognitive Ergonomics* and *Cognition Technology and Work*, and he sits on the Editorial Board of the *International Journal of Applied Aviation Studies*. In association with Neville Stanton (Southampton University) and Eduardo Salas (University of Central Florida) he is Series Editor for the Ashgate series of research monographs *Human Factors in Defence*.

Don regularly gives seminars to early career researchers working in industry, academia and research establishments on the preparation and submission of high quality manuscripts to Human Factors journals.

Preface

This book is the product of a series of one day courses that I have run over the years, delivered specifically for the benefit of authors early in their research career and new to the preparation of journal manuscripts. It contains the things that I wish my mentors and supervisors had told me 25 years ago (but didn't). Basically, I had to spend nearly quarter of a century finding all these things out for myself, usually by trial and error (but mostly error)!

When I became a manuscript reviewer, and ultimately a journal editor, I began to develop a further perspective on the publication process, one that is hidden to most early career authors. Not only did I become more critical of the manuscripts submitted, I also began to understand what the journal itself required (and why). As a writer, there are great benefits from having this additional perspective. It is one thing to *write* a good scientific paper, however you need to get the paper *published*. This requires some additional nous.

As an editor, you get to see a great many manuscripts. Time after time you see the same mistakes being made. Manuscripts are rejected or sent back for major re-work not because the science is bad, but because the same problems keep occurring in the way that the material is presented. And most of these issues are easy things to fix.

I have tried to make this book into a manual for constructing a journal manuscript: read a chapter – write a section. I don't claim that this is the best way to write a journal paper, however I have found that it works for me and for many of the people to whom I have described this technique. Use all content judiciously. Ultimately, though, you must find your own way of doing it. It is also worth persevering with this book – when you get to the end you might even find a tip to help you pay for your purchase!

I anticipate many readers will not agree with a great deal of what I say, and to an extent every now and then I do metaphorically poke them with a pointed stick. But I always remember what Robert Elwell, an old friend and a past book review editor for *Ergonomics*, once told me. He pointed out that when you read *anything* you connect with it on an emotional level first, before making any rational appraisal of its content. This includes text books and journal papers. Do you like it? For a book to work the reader must *engage* with it. Even if they are disagreeing with what is said, the reader is interacting with the words (and this is why Sid Dekker's work can be appealing to many). The words on a page must never become just wallpaper, otherwise what is the point?

I hope that you find the following engaging and useful.

Endorsements

The most difficult challenge facing early career researchers is making the transition from researcher to published researcher. In what I can honestly say is one of the most useful books I have ever read, Don Harris provides detailed step-by-step guidance for researchers wishing to publish their work in peer-reviewed academic journals. Just about everything budding authors need to know is covered, ranging from impact factors and citation rates to specific guidance on structure, writing style, and responding to reviewers comments. With the threat to "publish or perish" looming as large as ever, this book makes a timely and essential addition to the Human Factors catalogue. Written in Don's own unique style, it is a joy to read, and is a must have for students, early career researchers, and even experienced academics wishing to enhance academic outputs. Buy it now and become prolific...you will thank Don later.

Paul Salmon, Monash University, Australia

As a reviewer of countless scientific manuscripts over the years, I have to tell you that this book is sorely needed! If you only knew how many solid studies never see the light of day because the authors were unable to tell their story you would be shocked. This is a must-have, "how-to" book on writing up your research, kind of "journal-paper writing for dummies" – perhaps that's why I liked it so much. In writing this soon to be best-selling book, Don has struck an enviable compromise between breadth and depth, and I especially like the conversational tone throughout. In short, this book is intended to be used, not just a bookcase ornament; I can envision many pages with yellow sticky notes and rabbit-eared corners.

Scott A. Shappell, Clemson University, USA

A comprehensive and easily accessible guide, nicely written by somebody who has been at the receiving end of much scientific human factors research. Don Harris's Writing Human Factors Research Papers *is full of tips and examples of what to do, and what not to do. A much needed contribution to all human factors research training.*

Sidney Dekker, Griffith University, Australia

Chapter 1
Before You Start (Writing...)

Although it is a little unusual, it is necessary to make it clear right from the start what this book *is not* about. This book is not about how to do good Human Factors research, nor is it about how to analyse such research. It is assumed that both of these things are already in place. What this book is concerned with is how to prepare successful research papers of a quality fit for publication in an academic journal. If you do not have any high quality research upon which to base your paper (or papers) then not even the most accomplished author in the world will be able to produce a manuscript fit for printing in a journal. There is an old saying: 'you can't polish a turd'. Unfortunately, though, some potential authors believe that if you roll it in glitter this will be enough to assure its publication. It won't...

In over 25 years of refereeing Human Factors manuscripts and as a journal editor, the majority of the manuscripts that I have seen rejected have not been declined on the basis of the basic quality of the scientific work in them. Although in most of these cases the work has not been Earth shattering, it has been more than satisfactory. The reason why many manuscripts are rejected is simply because their content is not presented in a manner suitable for publication. Put another way, many authors with one kiss of their pen, can successfully turn Princes (or Princesses) into frogs. This book aims to set out one method by which authors will be able to present the story describing their research in such a manner that will satisfy most journal reviewers and editors. And it should be emphasised that writing a research paper is (in many ways) telling a story. Stories are good to read, and like a story, a research paper should have a beginning, a middle and an end, and it should progress logically. However, unlike most good novels there should be no surprises at its conclusion.

In the following chapters I hope to set out a recipe that any first time writer of a Human Factors paper can follow. Not every suggestion or sub-heading is applicable to every manuscript (use all content judiciously). However, by working through each chapter in the book and assessing your progress against the checklist at the end, a half-decent paper should emerge. This is very much writing by numbers, but it is a method that has worked for me and has proven to be successful. I am, however, still waiting to receive my first Nobel Prize.

Why Should You Write Papers?

There is no one answer to this question. However for any scientist, writing high quality journal papers can only be considered to be 'a good thing'. If you intend to make a career in the university or research sector then journal papers are essential to ensure career progression; it is part of your job. As a post-graduate student, an emerging publication record will considerably aid you in finding your first professional post.

If you are currently undertaking a PhD *don't* start writing papers after you have completed your thesis (despite what your supervisor may say). Write them while you are doing your research. Many PhD theses consist of at least two inter-related studies. Write up the first study while you are doing the second study. After you have submitted your manuscript you will receive comments back from the referees concerning all the aspects of your work (literature covered; analysis of results; their interpretation; presentation of the work, etc.). These are useful comments to incorporate into your thesis – think of them as corrections avoided. Furthermore, you get feedback (for free) on the adequacy of your work from two (or more) international experts in your particular area of Human Factors. And, if/when you get your paper accepted, it is worth listing the published papers derived from the work at the front of your thesis. This will make it very hard for any examiner to fail it on the basis of the scientific quality of the work. However, this is not to say that only papers derived from PhD studies are of the appropriate quality to publish in academic journals. Many Masters students that I have supervised have also had their work published.

From the viewpoint of any academic department, publishing papers in high quality journals is now essential. University (and hence departmental income) is dependent upon the grading of research quality (for example, as assessed in audits such as the UK University Research Assessment Exercise/Research Excellence Framework – see http://www.rae.ac.uk/ or http://www.hefce.ac.uk/research/ref/). Universities cannot afford to have research inactive members of staff at any level. This costs money. Attracting new students and future research income is based upon the reputation of the department, and the reputation of the department is a product of the people in it. The same is also true of governmental and private research laboratories.

If the work has been undertaken on a research or governmental grant there is often an expectation that it will lead to Journal papers. Not only do research councils like to see their funded work disseminated, it also gives these bodies confidence that the work they have underwritten is of appropriate scientific quality. The manuscript referees are effectively an independent audit source. Failure to get work published in these circumstances may adversely affect future grant applications. In short, a department cannot afford for its staff not to publish.

However, there are much, much more compelling reasons to publish your work. The first thing that usually comes out of any research programme is some kind of technical report for the sponsor (or a PhD thesis). This is then often consigned to 'shelfware'; that graveyard for many academic endeavours. But research is there to be used: publishing is a way that maximises its chances of being used by someone in the research community. Furthermore, seeing your research paper published makes you feel good. It is extremely satisfying to see your name in print (and it still is, even 25 years later)! The feeling never goes away no matter how many papers you publish. The only feeling better than holding your first paper in your hand (none of this electronic rubbish...) is seeing your name on the spine of a book. In the ranks of academia nothing can replace a little self-massaging of the ego. After all, no one else will do it for you.

Some Examples

Throughout the book I continually draw upon illustrations from several papers (on which I was one of the authors) that have been successfully published. Some complete papers (with commentary) are contained in the Appendices. I do not use my own work simply out of vanity. However, in this way throughout the book I can illustrate my thinking and the writing processes behind the papers that I have had a hand in constructing. I don't claim that my approach is the very best way of doing it – I do know that it is good enough to get published, though. Furthermore, by utilising examples and extracts from some of my own past papers I also don't have to pay publishers for the rights to reproduce them.

In Appendix 1 I have included a paper that I wrote with a colleague over a decade ago (Harris and Maxwell, 2001: 'Some Considerations for the Development of Effective Countermeasures to Aircrew use of Alcohol While Flying', which was published in the *International Journal of Aviation Psychology*). This work illustrates some of the problems and (my) solutions to the reporting of survey-based research.

Structure is everything when trying to get your message across. In Appendix 2 I have included a paper which is a little unusual as it has three parts to it, each written up as a mini-study (Harris, Chan-Pensley and McGarry, 2005, previously published in *Ergonomics*). Doing it this way helped to simplify and clarify the message in the paper both for the writer and the reader. Doing simple things to structure the paper, such as providing appropriate headings and sub-headings, also helps to guide the reader.

Both of the previous papers are survey-based works which use a number of tables and figures to summarise the data and which also contain a range of fairly complex statistical analyses. They were chosen deliberately for exactly these reasons. Huddlestone and Harris (2007), included in Appendix 3 (previously published in *Human Factors and Aerospace Safety*) is a completely different beast in that it uses qualitative data analysed without the use of numeric, statistical analyses. The reporting of qualitative research poses very different challenges to those faced by authors describing a study utilising numerical data.

Finally, Appendix 4 (Demagalski, Harris and Gautrey, 2002, also published in *Human Factors and Aerospace Safety*) contains a paper that employed an experimental approach for data collection, albeit one that used an engineering flight simulator and a number of complex flight tasks. The data are analysed using fairly conventional statistical analyses.

It is hoped that these four papers contain a reasonably wide range of approaches to illustrate the issues discussed in this book. Extracts from other papers are also included, as needed. These also provide a bit of variety. Please feel free to download copies of these from the publishers. If you get to the end of this book you will see why.

Have You Got Something to Say?

If you have got this far then it is safe to assume that you still have the desire to publish something. But have you got anything to say?

One of the most common excuses I used to hear for not publishing (apart from 'I'm too busy') was that the person didn't have anything interesting to write about. But there is usually something useful to say in any piece of research. What these people usually meant was either 'I don't *think* that I have enough to say' or 'I don't *think* that I have anything interesting to say'. They were usually wrong on at least one of these accounts.

One of the biggest mistakes when writing any paper is trying to say too much. The starting point for any journal paper is the results contained in it. Just select one or two results from your work that are theoretically (and maybe practically) interesting and which taken together make a relatively simple argument. If you want to tell a larger story, then consider releasing your work as a series of inter-related papers (or maybe a two-part paper if it is not too long).

Your 'Big Message' and Your Story

This introduces the first important points in writing any good scientific journal paper. To start off with you need a message and a story to tell. Your first consideration should be *what is the 'Big Message' in your paper?* Writing a journal paper is a little like writing a story in a newspaper: it has to have an 'angle' to it. 'Dog bites man': not interesting. 'Man bites dog' might be interesting, but you never read about the aeroplane that didn't crash or the politician's legitimate expenses claims. However, in this case you need a scientific angle. It need not be revolutionary but in some way it has to be new.

To illustrate, the 'Big Message' in your paper should be simple and easy to understand. It should implicitly run right the way through the manuscript. In Harris and Maxwell (2001) this 'Big Message' can simply be summarised as:

- You need to use the right countermeasures to reduce the likelihood of drinking and flying otherwise all your efforts will go to waste – different people drink and fly for different reasons.

From this point you then need to reduce the whole story that will run throughout the paper to just five or six bullet points. In this drinking and flying paper these were:

- Drinking and flying is undesirable – new regulations are on the horizon (at the time of writing).
- A great deal is known from the development of drinking and driving countermeasures but these have never been applied to aviation.
- Effective countermeasures depend upon the reasons for the underlying offending behaviour.
- A survey of 400+ pilots identified four theoretically recognisable factors for drinking and flying countermeasures: Education; Enforcement; Counselling; Sanctions.
- Different countermeasures were rated as being more likely to be effective in different groups of pilot (defined by licence type and drink/flying category).
- The results support specific deterrence theory (this was the main scientific 'hook').

My preferred *modus operandi* from this point is to print out these bullet points in a large font and pin them to the wall above the desk where I work. From this point *everything* that I write should help to tell the story encapsulated in them. Equally importantly, there should be nothing in the manuscript that *does not* follow directly from these notes. Brevity is a virtue in academic writing.

You have to remember that when writing for a scientific journal the manuscript *must* have a strong theoretical basis to it even if it is a paper with heavy emphasis on the practical application of scientific principles. Papers where the overriding emphasis is on practical application may be acceptable to some journals as a practitioner (rather than a scientific) paper.

Writing for Your Reader

Throughout the book I will constantly (perhaps irritatingly) keep referring to 'the reader' or 'your reader'. This is completely intentional. You always write for your reader and so you must consistently put yourself in the place of this fictitious person.

You can assume that your reader is an intelligent reader, with a good grasp of the fundamental issues and methods underpinning the science of Human Factors. If you are writing for a Journal focussed on research in a particular domain (e.g. Defence or Aerospace) you can probably also assume at least a basic knowledge of the application area. However, while it is important not to 'talk down' to your reader, don't expect too much of them. With the possible exception of your referees, most readers will be nowhere near as familiar as you with the issues in your research. A *brief* explanation of a few key issues can often go a long way to promoting understanding.

The first readers that you need to be aware of when writing are the manuscript referees. If you can't satisfy these readers then you don't need to worry about any others. The referees are critical readers who want to ensure the scientific integrity of your paper and ensure that it is clear and concise. While *at first* most referees' comments appear to be an attempt to torpedo your hard work and ensure that it never sees the light of day, a good referee will try to be constructive in their criticism and suggest revisions to improve your work (there are bad referees out there, though)!

Your other (eventual) readers are the members of the scientific community. These people will be reading your paper for a reason, either to better inform themselves about a particular aspect of Human Factors or to extract specific information (e.g. from the methodology or the findings from your work). Think about how and why you use a published scientific paper and extract information from it. There are rarely 'neutral' readers, reading simply for interest. In all cases, though, you should not be afraid of subtly 'selling' the contents of your paper to them. Accentuate the positive and eliminate the negative (well, don't draw too much attention to it...).

Finally, the scientific community is an international community. Remember that many of your readers may not speak English as their first language. Try to write clearly, concisely and avoid unnecessarily obscure words and expressions. It's science, not Shakespeare.

An Overview of What is to Follow

Writing a scientific paper is a bit like constructing a crime thriller (so I am told). When composing a murder mystery the first thing that you need to do is decide who killed who, where and using what sort of murder weapon (a bit like Cluedo™). Then you need to make sure that bits of evidence are left around for your intrepid detective to find. Having done all this you can then construct the story about how the book's main character finds all the clues to come to the same conclusion as you (the author). Note that you do not start at the beginning and work your way to the end of the story. The way to do it is the complete reverse of this.

Writing a scientific paper commences with having something to say, which is a direct product of having an interesting result. In the method described in the following pages and outlined in Figure 1.1, the paper writing process starts with the Results section. This is then followed by developing the Method section, which contains an explicit description of the manner by which the data described and analysed in the Results section were collected. Next comes the first and last bits of the Introduction, containing the *raison d'être* of the paper and its aims and objectives. Writing the Discussion section then follows this, interpreting the results within a wider practical and theoretical context, and at the end of this section there should be some nice strong Conclusions (relating directly to the aims and objectives of the work) and some Recommendations for future developments. Finally, the rest of the Introduction is developed, introducing the theoretical and practical background within which the Results will be discussed.

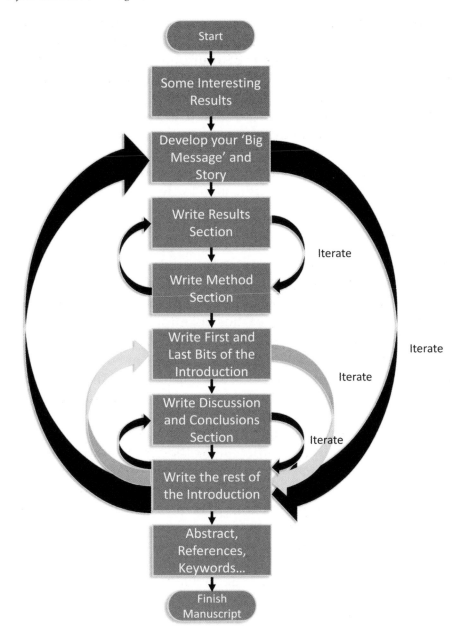

Figure 1.1   Overview of the manuscript writing process described in this book.

And just when you thought that you had finished various other bits and pieces then have to be done (Abstract; Title; References; Formatting; Keywords, etc.). You will have to 'iterate' between all the various parts several times (that's why there are all those arrows in Figure 1.1). A paper is far more than the sum of its component sections. It has to read as a 'whole'.

Constructing a manuscript in this way, section by section, ensures that everything that should be included is included. It also minimises the thought that is required (to a degree); you are just following a recipe. It is also far less daunting to do it in bite-sized chunks and the writing process can easily be fitted in around other work. However, when adopting this 'back to front' approach to writing a paper it is vital you bear in mind that the reader will start at the beginning and finish at the end and that the first two (very) critical readers, will be the referees.

At this stage of the process do not get too concerned about the formatting requirements for particular journals (references; callouts; section numbering conventions, etc.). The first requirement is to produce a comprehensive, coherent manuscript with a good story and a strong scientific basis underpinning it. All the formatting can all be done later.

Unfortunately, writing the manuscript is only half the battle. It is also the most interesting part. The second part of getting your paper published is submitting it to the target journal, getting the referees' comments back, responding to them and then re-submitting the manuscript. This process is summarised in Figure 1.2. Even then, assuming that the paper is accepted, you are not finished. Galley proofs need to be read, amended and returned to the publisher, along with other paperwork such as copyright clearances. After all this, the paper should be published but even then you may have to wait for up to a year to see your work in print (hard copy). Getting your manuscript published within a year of starting the paper writing process is a very good result. Expect a minimum of 18 months and do not be too surprised if it takes over two years. Many journals publish the dates a manuscript was first submitted, when revisions were received and when it was published. These will give you an idea of the average publication delay.

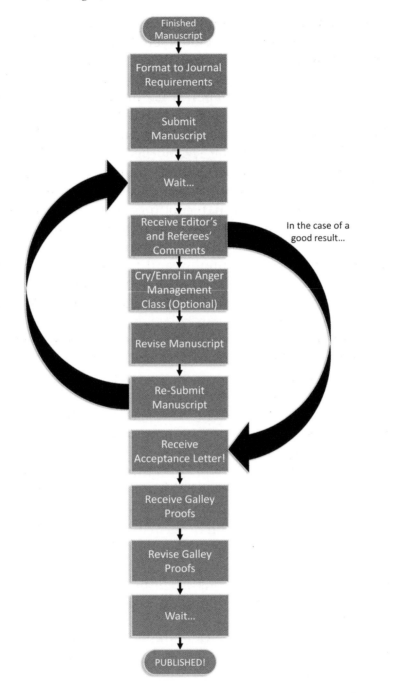

Figure 1.2   Overview of the manuscript submission, revision and re-submission process.

Prolific authors tend to have a number of manuscripts in the system at any one time, all at different stages of the publication process. However, most journals are now available on-line as well as in a printed form. Some of these journals offer an on-line first facility for papers that have completed the refereeing and production process but have not yet been allocated to a printed volume/edition of the publication (e.g. Taylor & Francis i*First*™). This content includes the paper's final DOI® (Digital Object Identifier) which is a unique reference number used for managing content on the internet. Publishing in this way can cut down the time before the work is available to the scientific community, but is it really 'published'? The jury is still out on this one.

Where Are You Going to Say It?

You may have noticed that I have not yet mentioned selecting a journal to whom to submit your manuscript. This should only be done after you have worked out what the 'Big Message' is in your paper and how you are going to present the story behind your research. The overall guiding principle, though, is don't try and publish work in a journal that's not interested.

There are many different Human Factors and Applied Psychological journals out there (I have no intention of listing them all). Each addresses a particular niche. You need to be aware of this. Some manuscripts submitted do not even get to the referees because they get a 'desk rejection' from the Editor. The most common reason for this is that the work within the paper is not within the scope of that particular publication. The first thing you need to check is that if similar work to yours (in terms of the methodological approach, the scientific content and/or the application area) has been published in the journal that you are considering. Check back through the last five years or so (this is easily done on-line). If the journal in question does not have a history of accepting papers like yours then either: (a) forget it, or (b) check it out with the Editor (just a quick e-mail) to see if they would consider it.

There are many other considerations that you need to make before deciding upon your final selection. If you work for a university then you will always need to worship at the altar of the Great Gods that are the research quality audits (such as the aforementioned Research Assessment Exercise/Research Excellence Framework). It is likely that future versions of these will be (to some extent) dependent upon bibliometric parameters which reflect the 'quality' of the papers published and the 'quality' of the journals in which they are published.

The most important parameter for these purposes will probably be the citation rate for a paper. However, this is impossible to predict when you are actually preparing the manuscript, although as we will see later, there are things that you may be able to do to increase the likelihood of your work being cited after it has been published. However, it is likely that you will be strongly encouraged to publish your work in a 'good' journal. Probably the key bibliometric parameter in this respect is the journal's Impact Factor (IF). This is simply the average number of citations received per paper published in that journal during the two preceding years (so if a journal has an IF of 2.00, then its papers published in the two years immediately prior to this received, on average, two citations each).

There are other bibliometric measures of the quality of a journal. The Immediacy Index is an indication of how quickly, on average, a paper in a journal is cited in other papers. It shows how often papers in a journal are cited within the same year (derived by dividing the number of citations to papers published in a particular year by the number of papers published in the same year). The Journal Citing Half-Life is the number of years from a given year that account for 50% of the citations published by a journal in its references. This gives an indication of the age of the majority of papers referenced by a journal. For a fuller description of these bibliometric indices try http://thomsonreuters.com/products_services/science/academic/. However, all of these indices should be taken with a very large pinch of salt. You can find many criticisms of this bibliometric approach to assessing journal quality on-line. It is not the intent of this book to discuss their pros and cons. The key point here is that if the journal in which you are considering publishing does not have an IF (or Immediacy Index or Citing Half-Life) then it may be institutional policy not to submit papers to such a periodical. It doesn't make it a bad journal, though. It may just be a new journal (it takes up to five years to qualify for an IF).

If there is pressure to get the work published in (relatively) short order, then it is worth considering one of the journals that appears more frequently than just quarterly (some journals appear six, ten or even 12 times per year). However, this may only speed up the process marginally. Speed of publication depends upon many things, and there is a simple reason why these journals are published more than just four times per year – they get a lot of papers. Keep an eye open for special issues of journals that may occur in your area of interest. This can help speed up the publication process and it also guarantees that your work will fit within the remit of the journal.

There are general Human Factors Journals (e.g. *Ergonomics; Applied Ergonomics; Cognition, Technology and Work* or *Human Factors*) and journals that serve a particular niche (such as the *International Journal of Aviation Psychology; Aviation Psychology and Applied Human Factors; Transportation Research (Part F); Human Factors and Ergonomics in Manufacturing* or the *International Journal of Industrial Ergonomics*). There are periodicals dedicated to safety (e.g. *Accident Analysis and Prevention; Journal of Occupational Safety and Ergonomics; Safety Science* or *Journal of Safety Research*). If you are working in computing there are any number of HCI related publications (e.g. *International Journal of Human-Computer Studies; International Journal of Human-Computer Interaction; Universal Access in the Information Society; Information Technology and People* or *Behaviour and Information Technology*). If your research has a strong psychological component to it, for example if it concerned with organisational aspects of behaviour, personnel selection, well-being, training or team working, then you may want to consider one of the Applied Psychology journals, such as *Journal of Occupational and Organizational Psychology; European Journal of Work and Organizational Psychology; Work and Stress; Small Group Research* or the *Journal of Applied Psychology*. There are also publications dedicated to review papers, such as *Theoretical Issues in Ergonomics Science; Applied Psychology (An International Review)* or *Organizational Psychology Review* (see Interlude III later in this book). These are merely examples and suggestions for publication outlets. Please do not consider it to be anywhere close to a comprehensive list of all the Human Factors-related journals out there. The one thing that I will almost guarantee, though, is that by the time you read this book, this list will be out of date. The 'Web of Science' (available from the aforementioned web site http://thomsonreuters.com/ products_services/science/science_products/a-z/web_of_science/) contains a comprehensive (but not exhaustive) list of journals. Note that those with no IF will not be included.

However, you should not restrict yourself to just considering the obvious Human Factors publications (from this point on you will have to bear with me for a while as I lapse back into the area of specialisation I know best – Human Factors in Aviation). Many engineering and technology journals (e.g. *Journal of Systems Engineering* or *Aerospace Science and Technology*) encourage the submission of Human Factors-related studies, as do multidisciplinary journals such as the *Journal of the Royal Aeronautical Society* or the *Journal of Navigation*. Many medical journals also accept Human Factors studies (e.g. *Aviation, Space and Environmental Medicine* or *Anesthesiology*). Journals associated with the practice of design also welcome similar papers (e.g. *Design Studies*; *Engineering Design* or *Engineering Design and Technology*). However, if you are considering submitting your manuscript to one of these non-mainstream Human Factors outlets then it is worth checking very carefully that your work is within the scope of the publication. You also need to be aware that when publishing in a non-Human Factors journal, you are dealing with a slightly different readership. While you can still consider them to be intelligent readers, a little more explanation of some of the Human Factors issues that you would normally take for granted may be required. You also need to gently explain why the Human Factors issues that you are talking about in your paper are important in the context of the subject matter normally dealt with in this journal. Don't be condescending, though.

You may now be getting close to selecting a target journal but there are still a few things worth checking out. For example, if you are doing work that elicits qualitative data have you ever seen any qualitative studies published in your proposed journal? Even though very few journals will specifically say that they do not accept qualitative research, there is often a heavy bias toward empirical studies which may prove to be an obstacle in the reviewing process.

Check the Editorial Board of the journal. The members of the Board may influence what you are going to write to some extent. There is a reasonable chance that at least one of your potential referees will be from the Board so you may want to consider how you describe their work in your paper (if you cite them). Editors also like to see recent papers from their journal in a manuscript's references (it helps the IF), so are there any relevant papers from the journal you can cite? If so, don't hesitate to include them.

Finally, once you have selected your target go and familiarise yourself with the journal's style. *This is important!* By this I do not mean the format for references, headings and sub-headings, etc. (these issues are considered in a later chapter). Look at how the published papers are constructed. Go and read several papers recently published in the journal. Is there one author who publishes in the journal regularly? Look at how s/he constructs their papers – they are obviously successful in getting their manuscripts published there. Some journals like long, comprehensive introductions; others prefer the introduction to be very short and punchy. Where does the journal require a description of the sample – in the Method or in the following Results section? Is there a word limit and/or a limit on the number of references cited? Some journals also limit the number of figures and tables. All of these things will help shape the way that you write and present your manuscript. In general, any manuscript *guidelines* published in the journal should be treated as *rules*. And remember, you need to play by the journal's rules. You are not doing them a favour by submitting your work to them. You should be trying to do everything that you can to make it as easy as possible for the Editor and the referees to accept your paper. Don't present them with a plethora of reasons to reject it, all set out on a bone china plate.

Finally, make your paper interesting and engaging. Even in scientific writing this is not a crime. People should want to read your work.[1] The mark of a 'good' paper is as much emotional as it is rational. Both scientists and your referees are human!

----

1   You may need to reconsider this paragraph if you read the material contained in the appendices.

Before You Start (Writing...): End of Chapter Checklist

Have you identified just a few results of interest that will form a coherent paper?	☐
Have you identified a 'Big Message' for your paper?	☐
Have you set out the story of your paper in a series of 5-6 simple bullet points?	☐
Does your paper have a theoretical basis?	☐
Have you identified a target journal to which to submit your manuscript?	☐
Have you familiarised yourself with the journal's style?	☐

OK – enough pointless idioms and mixed metaphors (but necessity is the mother of strange bedfellows). It's time to start writing. At this point just write in the way that you feel most comfortable. Get the words onto the page. Don't worry about any aspects of formatting at this stage. This will best addressed at a much later stage in the manuscript production process.

Interlude I:
A Few Short Observations on Structure and Style

The whole idea of this book is that it can almost be used as an instruction manual for constructing a Human Factors journal paper. Read a chapter: write a section. So, I figured that it would be best to put a few comments about structure and style at the beginning of the book before you write everything and then find out that you need to re-write it.

In this case I am not referring to the required journal structure and style for your paper. I am talking about good writing and presentational structure. Style is also a very idiosyncratic thing. Everyone develops their own manner of expression and all readers grow to love or hate the style of particular authors. I have no intention of suggesting a style to adopt. However, there are some things that are useful to bear in mind when preparing a manuscript.

Structure

There are golden rules for good structure. Firstly, start from the general and work to the specific. Paint the big picture first before filling in the details. This helps the reader to build a framework. You would not describe a house by describing the precise relationships between every brick and wooden frame and then expecting the reader to build up a mental model of all these connections in their head. Don't do this when you write.

There is an old adage when making presentations which applies equally to writing good papers:

- Tell them what you are going to tell them.
- Tell them it.
- Tell them what you told them.

Using descriptive headings and sub-headings helps to do this. Always remember that your readers are human. They do not have perfect memories for everything you have written; they may not be as familiar with the subject matter as you are; and they may not read your paper all at one sitting – they might get up and make a cup of coffee at some point. Good structure using headings and sub-headings to signpost the work allows the reader to do this. The occasional summary at an appropriate point can act as a well-placed *aide-memoir* (but don't overdo it – unnecessary repetition is a bad thing and can be irritating).

And one further point that I will continually make throughout this book. Tell the reader *why* they are being told *what* they are being told. *Never* let them make their own mind up. Draw all necessary conclusions for them on their behalf (but do it subtly).

Style

Improving just two basic aspects of style can avoid a great deal of irritation when reading a manuscript. Getting the tenses right (and consistent) and using short, simple sentences.

Tenses

It is traditional that all scientific writing is undertaken in the third person. This is supposed to reflect the impartiality of the scientist reporting the work. No person ever does anything: everything happens as if by magic. So never use *'I'* or *'we'*, etc. Few authors have a problem with this (even though it can occasionally lead to clumsy sentence construction). However, for some, tenses can be much more problematic.

There are some simple rules for being logical and consistent in the use of tenses in a manuscript. First of all, to assist in getting things right, imagine you are having a conversation with your reader about the research you are describing. I find that by doing this you naturally use the 'right' tense for the particular situation.

When you are describing the work of other researchers previously published (usually in the Introduction and also in the Discussion) you are telling the reader what they *did* (past tense). Their work was done some time ago, in the past.

> Norström (1978, 1981), Berger and Snortum (1986) and Albery and Guppy (1995) all **identified** a lack of a moral attachment to the law as a key determinant of drink-driving behavior. (Harris and Maxwell, 2001, p. 238 [emphasis added])

However, if you are pulling together or commenting on the findings from a number of authors (analysis and synthesis) and drawing your own conclusions from their work, you can imagine telling the reader about this in the present tense. The follow on sentence from the previous extract continues:

> Thus, effective countermeasures for non-believers **are** likely to be different to those of the inadvertent drink-flyer group as the root cause of the offending behaviour **is** different. (Harris and Maxwell, 2001, pp. 238–239 [emphasis added])

If in doubt, speak it out loud, as if talking to someone. As noted earlier, you will find that you use tenses quite naturally and in a similar manner to the above. This also helps to avoid over long, 'clumsy' sentences (see the following sub-section).

The Method section is very simple. You are telling the reader about something that you *did* to collect the data in the past (*past tense*).

The Results section is often best presented mostly (but not solely) in the past tense. You can imagine that you are telling the reader about the results of the analyses that you performed earlier which described the performance (in the *past*) of the participants during your trials:

> Pilots also **landed** significantly closer to the aiming point of the runway when using the new display system (F=6.57; df=1,8; p<0.05). Trial order, however, **had** no significant effect on this measure (F=0.29; df=1,8; p.>.05). (Demagalski, Harris and Gautrey, 2002, p. 186 [emphasis added])

However, when drawing the reader's attention to a particular result, the present tense is the one that fits most naturally:

> From figure 3 it **can** be seen that when using the emergency flight control display system only one 'landing' exceeded a 2,000 feet per minute rate of descent. (Demagalski, Harris and Gautrey, 2002, p. 187 [emphasis added])

The rules for use of tenses in the Discussion section are almost the same as those used in the Introduction. When describing the earlier work of other researchers use the *past* tense. When describing the work presented in the Results section, use the *past* tense. However, when making inferences and explanations based upon the work presented in your manuscript, use the *present* tense.

> Inspection of the inter-correlations between the dimensions, however, does suggest that drivers **can** distinguish between various aspects of their vehicles' dynamic behaviour and make separate evaluations of them. For example, the majority of the correlations involving the dimensions of 'ride comfort' and 'performance' **are** all relatively small (see table 4). (Harris, Chan-Pensley and McGarry, 2005, p. 793 [emphasis added])

These are by no means 'hard and fast' rules, and I suspect that many authors would disagree. However, they are good enough guidelines to ensure consistency. If you are uncertain, as I said earlier, read what you have written out loud. This works much better than reading it in your head for some reason. Awkward construction of tenses and phraseology becomes much more obvious. Which leads neatly to the next point.

Sentence Length

Keep sentences as short as possible. Whenever you use the word 'and' it is worth considering if you could split it into two sentences. I use a simple rule, which works for me. As I type this manuscript now (but not as it appears in the book you are reading) there are usually between 12–16 words per line. If a sentence is getting towards three lines in length, then I start to worry that it is too long. Once a sentence is over four lines long, then *it is* too long and will need cutting in two.

Try to avoid too many levels of embedded clause. Just one is enough. If you want to be psychological about it, too many embedded clauses overburden the Working Memory of your reader. You often have to read the sentence again to remind yourself how it started.

Finally, always remember that Human Factors is an international discipline and you are writing for many people who will not speak English as their first language. Don't be condescending but have a little consideration for them.

A Final Note on Style

A good writer writes with authority, even perhaps with the slightest touch of arrogance.

- Bad: 'Bloggs (1994) and Jones (1998) both observed a common, underlying tendency for sheep [ovis aries] to prefer consuming grass-based foodstuffs to lamb-based nutritional products, which it can be suggested is perhaps, largely for ideological reasons.'
- Good: 'Sheep prefer grass to lamb because they don't like eating their siblings (Bloggs, 1994; Jones, 1998).'[1]

Try to use the active voice as far as possible rather than the passive voice:

- Bad: 'The camel was licked by the boy'.
- Good: 'The boy licked the camel'.

Writing without equivocation helps to build the confidence in your reader that you know what you are talking about. It also makes for snappier, punchier sentences which have a clear point to them and makes your whole paper easier to follow. There are lots of jokes on web sites about common phrases found in psychological reports and what they actually mean:

- There is an underlying trend… (these data are practically meaningless).
- There is some suggestion that… (I think that).
- Anecdotal evidence suggests… (I can't find the reference).

Or my favourite on many academic's web pages:

- Selected publications… (all my publications).

These are all moderately amusing because they hide an underlying truth. Don't fall into the trap of writing in this manner. Say what you mean (or at least mean what you say).

---

1   This work is fictitious.

Chapter 2
Writing the Results Section

The Results section is the central component in any scientific paper. If you don't have good results then you don't have a paper. To re-iterate the point made in the opening chapter, you don't need a lot of results, just a few coherent results that can be clearly interpreted and linked to a strong theoretical explanation. Unfortunately, this will usually mean statistically significant results if you are writing an empirical paper, as there still remains a very heavy bias toward publishing only this type of finding. This is not to say that non-significant results are *insignificant* – it's just that they are rarely published. For the most part, this chapter will concentrate on presenting empirical data, however the reporting of qualitative data will be considered at the end. It will also be assumed that the data have been analysed correctly – this is a book about writing Human Factors academic papers, not statistical analysis.

In general, there are two distinct components in any Results section: description of the sample, and the presentation of the data and its analysis. Some journals may require the sample to be described in the preceding Method section, however you will have to do it somewhere and the same basic rules still apply. If appropriate, in another short sub-section you may also need to describe the manner in which your data were treated to compute the variables of ultimate interest prior to their analysis.

Papers that utilise qualitative data often combine the presentation of the Results with the Discussion section (the data, its analysis and its interpretation run together). Check this against exemplar papers from the journal to which you intend to submit your final manuscript.

Describing Your Sample

Irrespective of if your research is qualitative or quantitative in nature you must describe your sample thoroughly. This will usually be the first sub-section in the Results.

The reason for describing your sample is to demonstrate that the people who have contributed your data are a relevant and representative sample derived from the population of interest. Even at this point you need to start thinking about the Gestalt properties of your paper. When the reader looks at your sample description there should be no surprises. It should contain the sort of participants that they were expecting to see from (a) the material contained in the Introduction and (b) the description of the target sample characteristics in the Method section (see the following chapter). Remember this when you write these sections.

Ideally, the description of your sample should be done in prose rather than tabulated. As a golden rule, try and minimise the number of tables in a paper; put them in only where they are absolutely necessary. Tables are an excellent way of breaking down information and presenting it clearly (which is good) but they can take up a lot of physical page space. When producing a printed journal this is bad. Tables with only a few elements in them are an area on the printed page with a lot of white spaces and very few black marks containing information. Publishers (for printing and cost reasons) don't like this. This is why some journals put a limit on the number or tables (and figures). Paper costs money. You should keep one eye on the likely amount of physical space a final paper is likely to occupy. This is almost as important as being economical with the number of words that you use.

The description of your sample is also your chance to 'sell' your paper a bit. Tell the reader just how good the sample is. If you can, make it clear that it is a big sample. Note in the exemplar paper in Appendix 1, the first thing that the reader is told is that it is a *big* sample and that it is statistically representative.

> Four hundred and seventy-two completed survey instruments were returned. Nineteen further questionnaires were returned as being undeliverable, resulting in a final response rate of 48.1%. This ensured a probability of 0.95 that the sample obtained was within ± 0.1 SD of the true population mean (Hays, 1988). (Harris and Maxwell, 2001, pp. 241–242)

This also takes away ammunition from the referees. They cannot easily criticise the sample. However, because in this case it is quite a complex sample, some characteristics are described in the form of a table. However, more prosaic, but still necessary, information (male/female composition; mean age and flight experience) is presented in the form of prose.

If your sample isn't that 'good' in any respect, just describe it in very simple terms making no comment about its adequacy. *Do not* under any circumstances make any excuses or try to justify why the sample is small or unrepresentative, etc. This merely draws attention to the fact. In fact, this principle holds good throughout; don't draw attention to shortcomings in your paper by flagging them in this manner (unless you enjoy metaphorically shooting yourself in the foot). If these deficiencies are so bad, the referees will spot them anyway and point them out to you in their review.

Treatment of Data

The Method section describes the nature of the data collected and the means by which it was obtained. However in some instances the data that form the dependent variable(s) are subsequently computed from the measures initially collected. This sub-section in the manuscript describes how this is accomplished. Again, this part is sometimes found in the methodology section, but in common with the sample description, its content has to appear somewhere. Check the usual position of this sub-section in your target journal.

The following is taken from Rees and Harris (1995):

> The performance variables for participants for each approach [to the runway] were summarized to give two measures of error: the within-trial arithmetic mean error and the within-trial standard deviation of error… The arithmetic mean and the standard deviation are statistically independent. The former indicates any consistent error tendency. The latter describes the degree of within-subject variability in performance. Both have normally shaped distributions, making them suitable for the application of parametric statistics…

> Taken in combination, the within-trial arithmetic mean error and the within-trial standard deviation of error also completely define the root mean square error (RMSE). The RMSE also has the disadvantage in that it can produce identical values for quite disparate performances. For example, being consistently high, consistently low, or at the correct mean height but with great variations in height

keeping, may all result in the same root mean square value, which is taken to reflect tracking performance (Hubbard, 1987). As a result, it was considered that a combination of within-trial mean error and within-trial standard deviation of error best described participant's performance. (pp. 296–297)

Note that in this description of the treatment of the data to form the dependent variables, not only does it tell the reader *what* was done it also describes *why* it was done (further justified with reference to the work of a previous author). Such an explanation is often more neatly placed in this section rather than the Introduction. Put yourself in the position of the reader. When the reader is perusing the Introduction, they have little insight about what is going to happen several pages later. Don't ask them to keep in mind something not really connected to the flow of the other material in the Introduction.

The paper included in Appendix 1 (Harris and Maxwell, 2001) is unusual as there are two sub-sections on the treatment of the data. The first one describes how certain variables from the questionnaire were treated to form the groups of the independent variable; the following one describes how the data were manipulated to form the dependent variables. The level of detail provided in these sections should enable the competent reader to replicate the processes required to transform the data collected into the variables eventually subject to analysis. This same rule will apply in the following chapter concerning writing the Method section.

This sub-section also has other uses. For example, you may have decided to specifically exclude a sub-sample of your participants from the analyses; perhaps they were statistical outliers. The rationale for doing this and the parameter(s) used should be described and justified in this section (if you choose to admit to it). You may need to combine two (or more) smaller groups of the independent variable into a larger group, so that a meaningful comparison can be made. Again, the reason for this should be described and justified.

Presenting Your Results

Finally, the real 'meat' of the paper arrives. At this point in some way you will need to present the relationship between the levels of your independent variable(s) and the values of your dependent variable (I know that not all research uses dependent and independent variables but I am simply using these terms for brevity). This will usually involve either the tabulation of the results or their representation in some kind of figure; *do not use both!* Never repeat yourself in any way – it wastes space. A very common mistake in many manuscripts is to include a table of frequencies and then represent these data again as a histogram. Almost invariably, one of these will require deleting after the refereeing process has been completed. Remember this golden rule: minimise the number of tables and figures.

At this point it is a good idea to re-visit your 'Big Message'. Underlying any table or figure is a specific hypothesis which states that there is some sort of relationship between the variables represented. The tables or figures will help to present this relationship explicitly to the reader. Just check that the analyses you are presenting are in line with the aims and objectives of your paper. It is very easy to lose sight of what you are doing unless you constantly keep track of the bigger picture that you want to present.

Also with regards to your 'Big Message' what would be the best, most compelling way to represent your data to convey the point that you are trying to make? Is it graphically (e.g. to demonstrate a very strong linear trend between variables) or are your results too complex or abstract for that (e.g. a factor analysis of questionnaire data)? Figures can take up a lot of space, but if it is the best way to get your message across then use them. Always remember that something used sparingly has greater impact. Page after page of figures and diagrams can cease to be meaningful.

If your paper has several analyses in it, consider using sub-headings to structure the Results section (for example, see Section 5 in Harris, Chan-Pensley and McGarry, 2005 – Appendix 2). This helps to give the reader a frame of reference by introducing the reason for the analyses that follow.

Tables

The presentation of tables is a regular weakness in manuscripts. It is a fine balance between not wasting space and producing a table that is too dense and complex to extract information from it reliably. There are many different required formats for tables promulgated in the various publishers' guidelines (e.g. APA – American Psychological Association format) however they all have common elements in them and work toward the same simple aim: clarity. The guidelines are there to ensure that the content of the tables is unambiguous for the *reader* (note the emphasis) and it is easy to extract the required information from them. The guidelines are not intended to make author's lives a misery; however many authors make the reader's life miserable through the poor presentation of tables. Remember, the first readers that you are addressing are the manuscript referees. Ideally, you want to make sure the referees are on your side.

The most common weakness is that authors present fantastically complex tables with many nested elements to them using a very small font 'because they needed to fit it all in'. I would suggest that table 6 in Harris and Maxwell (2001) is about as dense (in terms of information) as you should go. Bigger tables are still OK (especially if they are presented in a landscape manner) however in terms of both the amount of information per square centimetre and the number of nested elements, table 6 is getting toward the limit for clarity. If you can't fit all the information in a single table then consider presenting it in two (or more) tables.

The format of the tables should reflect the hypotheses tested, so it follows that their layout and content must also compliment the statistical analyses performed. Note that in table 6 from Harris and Maxwell (2001) the first three columns and first three rows conform to the levels of two of the main effects in the Multivariate Analysis of Variance (MANOVA) performed (the third main effect is nested within the rows of the table). The main effects in such a statistical analysis reflect the overall effects (not broken down by treatment level – that is the interaction term) hence the inclusion of the row and column marginal means. Not only do these summary statistics reflect the nature of the statistical analysis, these figures will also aid later in the interpretation of the results in the Discussion section.

For the psychologists in the readership they might like to consider the Gestalt properties of the tables in Appendix 1. Note the emergent organisation in the tables as a product of the spacing of the rows, columns and fonts. This is not a banal comment. Most publishers' formatting guidelines require the use of a minimum number of horizontal lines to separate major table entries and no use of vertical lines. So, when preparing your tables you must allow adequate space between columns (and rows if not using a horizontal line) to promote readability.

Tables 2.1 and 2.2 are taken from a draft of the manuscript Demagalski, Harris and Gautrey (2002). I have chosen this example deliberately as it is quite a complex table. The version in Table 2.1 is horrible: this exemplifies everything that is bad: vertical and horizontal lines; use of computer package variable names; inappropriate number of decimal places; very crowded and unappealing to the eye. The same basic table presented in Table 2.2 is much better. Minimal use of horizontal lines and no vertical lines; columns spaced appropriately; self-explanatory row and column headings; data reported to an appropriate number of decimal places.

Table 2.1   Performance data taken from the draft manuscript of Demagalski, Harris and Gautrey (2002). Data are presented poorly.

DISPCON	ORD	MTR (SD)	MAD (SD)	NASATLX (SD)
COND1	1	102.6011 (14.9152)	879.0351 (202.3893)	42.7291 (16.2042)
	2	77.2973 (8.2348)	542.6244 (160.8733)	52.8776 (11.0243)
	TOT	89.972 (17.533)	710.838 (247.2)	47.857 (14.102)
COND2	1	326.4277 (132.9184)	1796.1491 (1842.8083)	70.3567 (8.6277)
	2	217.0238 (88.9354)	1139.9005 (1551.2158)	74.4781 (15.252)
	TOT	271.7596 (121.2782)	1468.0214 (1642.7487)	72.4863 (11.8568)

Table 2.2   The same performance data taken from the draft manuscript of Demagalski, Harris and Gautrey (2002). In this case data are presented in a typical format suitable for a manuscript to be submitted to an academic journal.

Display Condition	Trial Order	Mean time to recover (s.d.)	Mean absolute deviations (s.d.)	Overall NASA TLX score (s.d.)
Using emergency display system	1st	102.6 (14.9)	879.0 (202.3)	42.7 (16.2)
	2nd	77.2 (8.2)	542.6 (160.8)	52.8 (11.0)
	Overall	89.9 (17.5)	710.8 (247.2)	47.8 (14.1)
Without emergency display system	1st	326.4 (132.9)	1796.1 (1842.8)	70.3 (8.6)
	2nd	217.0 (88.9)	1139.9 (1551.2)	74.4 (15.2)
	Overall	271.7 (121.2)	1468.0 (1642.7)	72.4 (11.8)

It is worth taking time out to consider the units in which the variables are expressed. Most journals, both European and American, require the use of SI (*Système International*) units (metres, kilograms, seconds, etc.). The exception to this are the Aeronautical journals (especially those concerned with operational aspects) which may permit the use of that peculiar mix of units promulgated as an international standard by the ICAO (long distances in nautical miles; shorter distances in metres; altitude in feet; speed in knots; weight in kilograms... as they say 'go figure').

A personal bugbear is the effect that computer-based statistical analysis packages have had on the presentation of tables. It *is not* acceptable to use these tables directly in a manuscript unless they have been formatted correctly. *Do not* use variable names: these are not meaningful to anyone other than the analyst (see table 1). Including a glossary for the variable names *is not* an acceptable alternative. Row and column headings in the table should be self-explanatory. *Do not* report results to eight decimal places. Just because the computer calculates results to such precision does not mean it is meaningful to report them in this manner. If you use a five-point Likert scale, is it meaningful to report means to four decimal places? What does one ten-thousandth of a Likert scale point actually represent in the real world? Use common sense and a little discretion.

Finally, table headings should be descriptive and self-explanatory even to a reader who has not read the main text of the manuscript. Units of measure should be included (if appropriate). Table headings always go above tables. To illustrate, the final published caption for the exemplar table taken from Demagalski, Harris and Gautrey (2002), reads:

Table 1   Measures to assess performance when recovering the aircraft to straight and level flight with and without the emergency display system. The display was used in either the first or the second set of trials. Time to recover is measured in seconds; mean absolute deviations from assigned cruising altitude are measured in feet. (p. 184)

Normally, tables will be included either separately or at the end of the manuscript when it is finally submitted, with just a call-out in the text at an appropriate point. They are not included in the main text.

The reporting of simple mean values for a variable broken down over a limited number of groups should be done in the main text (not in a table). A typical format for doing this would be:

The mean age of Captains was 45.22 years (SD = 5.45). First Officers were typically younger (M = 33.45 years, SD = 3.45).

Never, ever under *any* circumstances report a mean without its associated standard deviation. Offenders should be forced to undergo ritualistic punishment involving a fresh pineapple; a wire brush and some small fish hooks.

Finally, a note of caution before preparing any table, but particularly a large one. Check the required format for the preparation of manuscripts, particularly the paper size (A4 or US letter); the required minimum margin size and the minimum font size. You don't want to spend a lot of time preparing a table only to find that it won't fit!

Figures

The presentation of figures is also a weakness in many manuscripts and in general, they share many common shortcomings with poorly-presented tables: they tend to be over-complex, contain too many elements and are often labelled in a very small font 'because the author needed to fit it all in'... Figures can take up a great deal of physical space on the printed page. It is often said that a picture paints a thousand words. This is not necessarily true. A well-drawn figure can make complex interrelationships and structures easy to understand but a poorly drawn figure can actually cause confusion.

Before deciding if you need to include a figure in your results section you should ask yourself if you really need it. The first test you should apply is deciding if the figure actually supports the 'Big Message' that you are trying to get across in your manuscript. Secondly, you should try and establish if your figure actually paints a thousand words, or is it going to take you a thousand words to explain your diagram? Thirdly, is it simply a waste of space? For example, a histogram depicting frequency counts in just two or three categories is better explained in a very simple table or in prose in the main text; another case of lots of white space on the page and little information. *Never* repeat information in a figure that has already been presented in a table.

If you decide that a figure of some kind is necessary follow the guidelines from the prescribed format for manuscripts for that particular journal very carefully. As I said earlier when discussing the format of tables, these guidelines are not promulgated to make your life a misery; they are there to ensure that the figures accompanying your manuscript will be legible when finally produced in the printed version of the journal. Treat these guidelines as rules. And, the emphasis is on preparing figures for the print version of the journal not the electronic format. This is for two reasons. Firstly there is a hard limit of the size of the printed page and secondly this format of the journal is still usually produced only in black and white.

As a strategy for preparing figures, first of all look at a printed version of the journal (*not* the electronic version or a print out of the electronic version). You need to first assess the maximum area available to you on the printed page. Printing out the web-based version of the paper onto an A4 page will not give you an accurate idea of the largest available dimensions for your figure. Often the limiting factor is not the number of lines or the size of the boxes in a diagram; it is the size of the labels. These must be legible when it is finally printed, so *do not* just keep making the font smaller so you can fit it all in.

Although all computer drawing packages can now produce 16 million different colours, printed journals are still (by and large) monochrome. This means that a great deal of care is required when using shading to depict different things. White, black and grey is OK: you *might* even get away with white, black and dark grey and light grey, but after that it is difficult to differentiate different shades of solid grey. You will have to start to use different fill patterns to distinguish between the various areas in your figure (see, for example, figure 2 in Harris, Chan-Pensley and McGarry, 2005, in Appendix 2). However, if you do this then make sure that the fill areas in any key to the figure are large enough for the reader to be able to distinguish between the different patterns.

Similarly, if different types of connecting lines have different meanings then the reader needs to be able to distinguish between then reliably. The only way reliable way of doing this is by using distinct patterns of dotted lines. Line colour or width will not do. It is also essential to include a key for these aspects of the figure as well (for example, see Figure 2.1, taken from Li, Harris and Yu, 2008).

One of the best tests that you can apply is to draw your figure (in black and white) and then reduce it on your computer or on a photocopier to the final size that it will actually be reproduced on the printed page. Can you read it? If the answer is 'no', then it will have to be simplified and re-drawn. When you finally submit the figure with your manuscript most journals will not accept output directly from a drawing package (e.g. PowerPoint®). This is because when re-sizing figures to fit the page, often labels will fail to re-size and/or change to the default font on that particular computer (you will probably have experienced this). The whole aspect ratio of the figure may also change if opened on a different page size from the one upon which it was created. As a result, after finalising your figure it will be necessary to convert it to

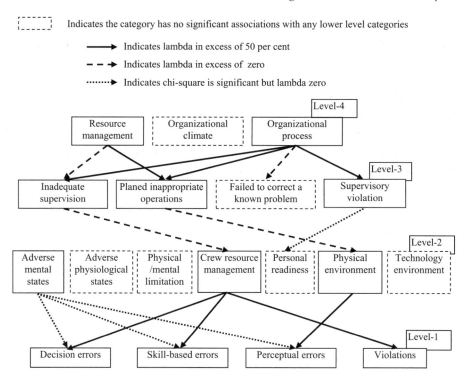

Figure 2.1   Exemplar figure (from Li, Harris and Yu, 2008). Note the different types of dotted line to indicate the different statistical categories of the various inter-relationships.

another format, usually Tagged Image File Format (TIFF file); Joint Photographic Experts Group (JPEG file); Graphics Interchange Format (GIF); or a device independent bitmap (BMP). These make re-sizing the figures much easier but check the quality of the final figure; it is often not as high as you would like. Furthermore, you cannot edit these files, so make sure that everything is spelt correctly before creating them. This includes ensuring the 'right' version of English is used (US or UK) for the target journal. Some journals now request that figures should be produced using drawing packages employing vector graphics (e.g. Adobe Illustrator®, Microsoft Visio® or Corel Draw®) so that they can be modified by the editorial staff. Re-sizing figures produced this way will not adversely affect their quality. Furthermore, when you prepare these files (a) make sure that you crop all the 'dead area' around your figure, otherwise you end up with a small figure in a sea of white space and (b) make sure you save them at the highest resolution possible (if they are a TIF; JPEG; GIF or BMP). When printed, the figures should reproduce at a minimum of 600 dpi (dots per inch).

With the dramatic improvements in digital imaging over the last ten years, journals will now also accept digital photographs. Make sure that these are of high quality and strictly relevant to the message in the paper. If applicable, the object of interest in the photograph should be shot against a plain background. If the images contain people, these should only be included with their consent and the subject in the photograph should be anonymised. Again, check that the photographs reproduce well in black and white.

The same rules apply for figure captions as table headings (however in this case they go below the figure). They should be descriptive and self-explanatory to a reader who has not read the main text of the manuscript. Units of measure should be included (if appropriate). While brevity of expression is something to be commended in any aspect of scientific writing it is perfectly acceptable for a figure caption to span over three or four lines in order to make it self-explanatory. For example, the caption to figure 2 in Harris, Chan-Pensley and McGarry (2005) reads:

Figure 2.   Example of the graphical presentation of the multidimensional vehicle dynamic qualities rating scale contrasting a poorly rated sports coupe with a highly rated executive saloon. NB: Lower ratings indicate more desirable dynamic behaviours and narrow histogram bars indicate higher levels of importance of a particular dimension with regard to that vehicle's market sector. To interpret a dynamic qualities profile it should be remembered that if a bar is narrow it should also be short. (p. 979)

Note that the previous exemplar caption for a table was also quite long for exactly the same reason.

In a similar manner to tables, figures will usually be included as separate files appended to the manuscript when it is finally submitted. Their approximate required position should be indicated as a call-out in the text at an appropriate point.

Reporting Results of Statistical Analyses

Another common effect that the use of computer-based statistics packages has had on the presentation of manuscripts is for authors to include the output tables to present the results of these analyses (for example full ANOVA tables or the complete tables resulting from regression analyses). These tables will include critical values and probabilities usually to at least four decimal places and will also include many statistics not needed and never subsequently referred to. This is also bad form. The results from statistical tests should be reported in the standard abbreviated format in the prose describing the content of the tables and their accompanying analyses. The following are examples of the general format for common statistical tests (taken from the APA):

- *Independent Groups t-test*: t(number of degrees of freedom) = t value, p value
    - *Example*: $t(16) = 3.34, p < 0.05$

- *Simple Independent Groups ANOVA (or for a main effect/ interaction term in an n-way ANOVA)*: F (degrees of freedom in numerator, denominator) = F value, p value
    - *Example*: $F(2, 50) = 9.35, p < 0.001$

  Any additional statistics (such as η^2 values or ω^2 values) should be placed after the probability value. *Post hoc* test should be contained in the prose that follow and refer specifically to the test used, the contrast being made, the value for the chosen statistical parameter and its associated probability value.

- *Correlation: r* (number of degrees of freedom) = r value, p value
    - *Example*: $r(55) = 0.49, p < 0.01$

- *Simple Linear Regression:* There are two components to any regression result; the regression equation and the significance of the regression equation (its fit to the data). Taking these in reverse order, the statistical test of the regression line is reported in exactly the same way as an ANOVA or a t-test (either statistic may be used). When reporting the equation of the line the R value, R^2 value and the slope of the line (either standardised and/or unstandardised) should also be included. In the case of the unstandardised equation, don't forget to include the intercept term. If the regression equation is accompanied by a figure, one option may be to include the equation of the line on the graph. However, figures describing a regression equation often occupy a great deal of space and add little explanatory power.

- *Multiple Regression:* The statistics associated with the overall solution should be reported in the same manner as for simple linear regression, however the individual regression weights for each of the independent (predictor) variables should also be included, as should their associated t-values and probability.

- χ^2 *test of association:* χ^2 (degrees of freedom, N = sample size) = χ^2 value, p value
  - *Example:* χ^2 (2, N= 524) = 9.35, p < 0.02

Opinions are now slightly divided on the correct way to present the probabilities associated with the critical value of a test statistic. Strictly speaking, the acceptance of the alternative hypothesis is based upon a decision made against a pre-specified probability value (a chosen alpha level), typically p = 0.05. The probability value when results are reported should reflect this decision criterion (hence the use of the '<' to indicate a value lower than the pre-specified alpha value). However most statistical packages now calculate an exact probability value associated with a test statistic, and some publishers now prefer that exact p values are reported. If a p-value is non-significant then the abbreviation 'ns' (non-significant) may be used or the calculated p-value may be reported prefaced by an equals sign.

Definitive guidance for reporting of more complex, multivariate statistics is much harder to provide. Reporting principal component or factor analyses will usually involve a tabulation of the factor loadings associated with each variable on each of the components/factors extracted (e.g. Harris and Maxwell, 2001, table 4). The summary statistics pertaining to each component will also need to be reported (see table 5 in the same paper). An almost identical format is used in tables 1 and 2 from Harris,

Chan-Pensley and McGarry (2005), however in this case, for clarity only the variables that load heavily on a particular factor (over 0.45) are included.

When reporting a confirmatory factor analysis a diagrammatic representation of the factor structure may be clearer, with the various loadings represented on the arrows in the figure. However, this will still need to be supplemented with other pertinent statistics describing inter-correlations between latent variables and the overall 'goodness of fit' of the solution to the hypothesised underlying structure (see Harris, Chan-Pensley and McGarry, 2005, section 4.2, figure 1).

Reporting the results from MANOVAs and Discriminant Function Analyses are always likely to involve many tables that relate to the statistical analysis (see table 7 in Harris and Maxwell, 2001). Even publications such as the APA publication manual are slightly loath to provide specific direction for reporting such analyses. I would suggest two 'rules of thumb' can be applied in these circumstances. Firstly, the analyses reported should relate to the appropriate tables of results (a point made in the earlier sub-section) and secondly, the any statistical parameters reported in the tables should be used in the interpretation of the results. Modern statistical packages will produce a plethora of parameters to describe the relationship between variables and their associated significance values. You don't have to report all of them. Just report the ones that are used. If in doubt, check for similar analyses previously published in your target journal. What parameters did these authors report?

As a final note, do not attempt to interpret the results of tables, figures or the statistical analyses in the Results section. This can wait until the Discussion. Any comment on the results should be fairly bland statements along the lines of '*Group A had a significantly higher mean score than Group B*' or '*there was a significant linear association between A and B*' (for example, see paragraph following tables 6 and 7 in Harris and Maxwell, 2001). Do not simply re-state the content of any table or graph in prose. Furthermore, when making these comments always explicitly refer to the table or figure (by number) where the summary data can be found.

Reporting the Results of Qualitative Analyses

Reporting the results from qualitative analyses in a succinct manner is quite a considerable challenge. There is no one, easy way of doing this. Often the Results and Discussion sections run together in papers using qualitative data so that the analysis and interpretation of the data within a wider theoretical context can sit side-by-side (see Chapter 5 for further explanation). This is often neater and allows for an easier discussion of the resulting issues. It is difficult to generalise, but in many cases the analysis of qualitative data may either be driven by some kind of existing theoretical model (as in Psymouli, Harris and Irving, 2005 – see later) or lead to the development of a model (for example, Huddlestone and Harris, 2007; Appendix 3).

One issue when reporting qualitative data has absolutely nothing to do with the quality of the data and its analysis, but has everything to do with the quality of the author's writing. It is vital to develop the reader's confidence that the analysis and interpretation of the results is sound. For the reader, this begins in the Method section through the painstaking description of the manner by which the data were elicited and subsequently analysed. The confidence building process continues in the Results section through the meticulous presentation of the data and their interpretation. This is much harder than presenting tables of numbers and their accompanying statistical analyses. Furthermore, remember that when interpreting qualitative data the emphasis is often upon providing explanation for phenomena and not on categorisation and quantifying their frequency of occurrence. Just a single observation or comment among many might be the key to the whole interpretation of the data obtained.

In common with reporting quantitative analyses, the Results section should still start with a description of the sample, however it is likely that in this case the sample will be smaller. Depending upon the objectives of the paper emphasis may be placed on the appropriateness and skills in the sample, demonstrating the validity of the data obtained. In certain circumstances it may even be appropriate to provide thumbnail sketches of individual participants to establish their credibility. For example:

Seven SMEs also took part. This group comprised an HSE (Health and Safety Executive) specialist inspector and expert on earth moving plant, the managing director and the head of engineering from a small dumper manufacturer, the head of engineering and senior design engineer from the UK's largest manufacturer of dumpers, a senior instructor for construction skills from the National Construction Industry Training Board, the managing director of a regional company providing plant training and an expert witness on dumper accidents. These SMEs were deliberately chosen on the basis that they were all regarded in some capacity as construction site dumper experts who could offer differing perspectives on the operation and hazards and risks associated with the on-site use of dumpers. (Bohm and Harris, 2010, p. 5)

Authors at this point should also provide other details of the data collected, for example the number of hours of interviews; the number of hours of observations made or the number of accident reports analysed and coded. Do not be afraid to stress just how much data you have collected (if you have collected a significant amount). This will help to establish the comprehensive coverage of the material obtained, hence its generalisability.

If the approach requires that the data be analysed within an extant model, that framework should itself be justified to be theoretically (and practically) legitimate:

Interview data, notes from the observations and the additional data were subject to content analysis using initially pre-determined categories (Ezzy, 2002). The coding framework was largely derived from the works of Lock and Strutt (1985) and Drury, Prabhu and Gramopadhye (1990) and modified in light of data gained during the observation of aircraft inspections to incorporate the additional steps required for the inspection of composite materials. This framework is described in the flowchart presented in figure 3. The emphasis in this paper, however, is placed firmly within the 'Search', 'Decision Making' and 'Respond' phases, as described in the framework published by Drury, Prabhu and Gramopadhye (1990). (Psymouli, Harris and Irving, 2005, pp. 96–97)

In instances where narrative or observational data have been re-coded and categorised, the next part of the Results should deal with issues surrounding the intra- and inter-rater reliability of this process. From the reader's perspective, the method of doing this should already have been described in the preceding major section. However, the results of this analysis should be presented. Again, this helps to build the reader's confidence about the reliability and validity of the data (and always remember that your first critical readers are going to be the referees for the journal)! The following description of the results relating to determining the inter-rater reliability from the coding of narrative aircraft incident data is taken from Li and Harris (2006).

> The inter-rater reliabilities calculated on a category-by-category basis were assessed using Cohen's Kappa. The values obtained ranged between 0.44 and 0.83 ... Fourteen HFACS categories exceeded a Kappa of 0.60 indicating substantial agreement [Landis & Koch, 1977]. As Cohen's Kappa can produce misleadingly low figures for inter-rater reliability where the sample size is small or where there is very high agreement between raters associated with a large proportion of cases falling into one category [Huddlestone, 2003], inter-rater reliabilities were also calculated as a simple percentage rate of agreement. These showed reliability figures between 72.3% and 96.4%, further indicating acceptable reliability between the raters. (Li and Harris, 2006, p. 1058)

Inter-rater reliability statistics will not be applicable in all cases. They will only be required where there is the coding of data into an extant theoretical framework, or where a categorisation scheme has been developed from the data collected and the membership of those categories is going to provide the basis for further analysis (even if this is as simple as a count of their relative frequencies). Some would suggest that this is a form of quantitative analysis but the people that would engage in this argument are also the kind who would debate with Thomas Aquinas about how many angels can dance of the head of a pin. Pointless...

If the qualitative data analysis is being driven with reference to an extant theoretical framework (already introduced in the Introduction) then each inference drawn should explicitly refer back to the particular aspect of the model under examination, providing evidence from quotes or observations.

> With regard to the flowchart presented in figure 1, the former question is presented as an issue in the 'decision making' step. The latter issue is addressed in the 'respond' step (see Drury, Prabhu and Gramopadhye, 1990)...

> ...It was observed that the inspectors undertake this 'signal conditioning' task in several ways once the visual inspection has identified a potential area of damage. Several inspectors used a further, different type of visual inspection technique to their initial assessment of the potential damage.

> '...when visual[ly]inspecting a surface that it is suspected to have dents, we have to stand back, at some distance from the surface and use the reflection of the natural light on the surface in order to ascertain whether any dents occurred on that surface. The reflection of the light onto the aircraft panels is a

very important factor in discovering dents and other marks on the structures'. (Psymouli, Harris and Irving, 2005, pp. 100–101)

It's all about establishing an audit trail, from data to analysis to inference. This is vital.

Qualitative data analysis is not just about categorisation, though; it can also be about developing models of explanation. This is typical of research that adopts a quasi-grounded-theory based analysis of data using methods based around the procedures and techniques described by Strauss and Corbin (1990). Figure 2.2 is taken from Bohm and Harris (in press). It portrays a composite mental model of the factors underlying the loss of control of a dumper. These were elicited from a series of in-depth interviews undertaken with construction site dumper-truck drivers. The mental model is presented in the form of an influence diagram.

Support for the components and inter-relationships in the model is drawn from interview data. It is essential to provide evidence for aspects of the model and to establish a verifiable audit trail. Once again, this is typically done through the use of quotes to exemplify the various features. For example:

> There was a high level of confusion (45%) about what effect being laden or unladen would have on maintaining control of the dumper if travelling at speed (Figure 4). One driver's comments highlighted this confusion:

> 'I would have thought having a load would be better – I suppose it depends on what you mean. Presumably it would be more stable and more centered [with a load] but once having lost control I would imagine the loss of control would be greater because of the load'. (Bohm and Harris, in press)

Note that in this case, no attempt at all was made to quantify the interview responses. Emphasis was completely upon developing a diagrammatic representation encompassing all the potential factors in dumper-drivers' mental models of accident causation to provide explanation.

The approach used by Huddlestone and Harris (2007, (see Appendix 3), structures the analysis around the 'open', 'axial' and 'selective' coding processes prescribed by Strauss and Corbin (1990) when undertaking grounded-theory research. This provides an explicit link to the methodology and emphasises that a rigorous approach has been used to produce the data presented

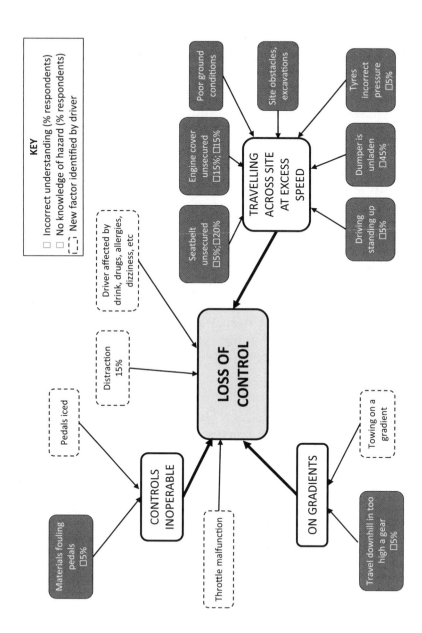

Figure 2.2 Composite SME (baseline) and driver mental model in the form of an influence diagram to describe the factors underlying the loss of control of a dumper.

in the Results. Note also that the Results section contains explicit references to the inter-rater reliability of the coding process to stress best practice has been adopted in the analysis of the qualitative data. All the time you should be looking to try to build the reader's confidence in your method, your data, your analysis and subsequently all the conclusions you draw. Again, the frequent use of appropriate quotes (selected to best illustrate the nature of the categories elicited) helps to establish an audit trail. Furthermore, they serve to bring the content of the section to life for the reader, linking the science to the real world context in which the data were collected. Such quotes can also bring a little 'light relief'. Reading any scientific paper should have a large element of interest, even pleasure, in it. It should not be a chore.

End of Chapter Summary

The key to effective written communication is for the writer to put themselves in the position of the reader. This comment applies to everything that you write. When you have finished preparing the results section, have a look at it and ask yourself if you would like to receive this to review? Is the section well organised; are the tables clear; are the figures legible; are the statistics reported properly?

You also need to re-visit you 'Big Message' and list of bullet points that contain your initial story. Do the data and the analyses that you have reported help to convey the bigger message in the paper? Are the analyses complete? More importantly, are there any data and analyses reported that make no contribution to the message in the paper; if so, why are they there? Good, well-focussed papers are as much a product of the material that you do not include, as they are the results of the things that you do include. Be ruthless!

Writing the Results Section: End of Chapter Checklist

Is there a clear and comprehensive description of the sample?	☐
If appropriate, have you described how the data were treated to form the dependent and independent variables used in the analyses?	☐
Are all your tables and figures necessary?	☐
Are all your tables and figures clear, and legible at the final size they will be printed in the Journal and when re-produced in black and white?	☐
Do your tables and figures have self-explanatory captions?	☐
Are your statistical results presented in the required format?	☐
If presenting qualitative data is there an explicit audit trail linking the inferences drawn to the source data?	☐
Does the content of the results section compliment the 'Big Message' and the 'story' as described in the summary bullet points?	☐
Have you removed any material that is not required to deliver the message within the paper?	☐
Would you want to read what you have just written?	☐

In the next section we will be looking at the relationship between the material in the Results section and the content of the Method section.

Chapter 3
Writing the Method Section

This is actually one of the easiest sections to write. With luck, it is almost all set out on a plate for you. All you need to do is fill in the blanks and make sure that it links up coherently with the material in the Results section and the subject matter in the Introduction. But remember, readers will not be reading the paper in the same order as you are writing it, so always keep an eye on your master plan. If in doubt, before writing go and remind yourself what your 'Big Message' is and what is in the series of bullet points that set out your story.

The basic material that needs to be described in the Method section has all been defined in the section you have just finished (the Results). In the Results section you will have included:

- A description of the sample.
- A list of Dependent and Independent variables.
- Potentially, a short description of how the data were transformed to make them into the variables analysed.
- An explicit formulation of the hypothesised relationship between these variables.

Basically, the Method section is all about describing how you collected these data from this target sample of people. It is amazing how many manuscripts submitted to journals contain detailed accounts of the process of collecting vast numbers of complex variables, many of which magically disappear (for no apparent reason) when the page is turned. If the variable isn't in the Results section then the reader doesn't need to know about it in the Method section.

There is an acid test that I apply to any Method section, irrespective of if I have written it or if I am reviewing the work of others. I call it the 'Delia Smith'[1] test. If you read a Delia Smith cookbook it begins with a list of ingredients which is followed by a simple, to the point, blow-by-blow description of how to assemble these ingredients into the final dish. Any half competent cook should be able to follow these instructions and produce exactly the same meal as Delia without having to resort to 'phoning her up for advice or e-mailing her (she is unlikely to answer, anyway). When you have finished writing the Method section you should ask yourself if any half-competent Human Factors researcher could follow your description and reproduce your work without having to get in touch with you. Have you explained yourself as clearly and concisely as Delia?

Not all of the following sections and sub-sections will apply to every paper. A great deal will depend upon the type of study (experimental; observational; survey, etc.) and nature of the data (quantitative or qualitative) produced. Pick and choose what you use, as required.

Description of Target Sample

The Method section often begins with a description of the target sample (e.g. size, demographic requirements and/or skill attributes needed) from whom to elicit data. In some cases, most often in the case of experimental work, this section is combined with the sample description in the Results. However, in survey-based research the sample that you aim for is often not quite the sample finally obtained. This section describes the desired characteristics of the sample and may also explain the manner by which the participants were recruited to the study (quota sampling; random sampling; opportunity sample, etc.). As an example, see the Sample sub-section in the Method section of Harris and Maxwell (2001) in Appendix 1, and contrast this with the subject matter of the Sample sub-section in the following Results section. There should be a great deal of correspondence between these two sub-sections but no unnecessary overlap. This part of the manuscript is about what you *did*, not what you *got*.

---

1   Insert any TV chef at this point according to your own particular tastes (e.g. Martha Stewart, Curtis Stone, Jamie Oliver, Michel Roux or my personal favourite, Keith Floyd).

This sub-section should also briefly describe how participants were recruited to the study (if applicable) and detail any remuneration that they received for taking part (e.g. monetary payments or course credits).

Description of Data Collection Equipment

The following sub-section describes the equipment or facilities used to collect the data. Descriptions of questionnaires or structured interview schedules should also be included in this part of the manuscript.

Equipment

The precise details of any commercially available equipment used (e.g. make and model; software name and version number) should be given. If it is not obvious, along with these details should be given a brief description of its function. Such particulars are unnecessary, however, for run-of-the-mill items, such as cassette/digital voice recorders or digital cameras.

If equipment has been specifically constructed for the study, then a short description should be provided. An overly detailed explanation is not required but there must be enough of an explanation to allow a clear insight into its arrangement and function. If necessary, the reader can be directed to an internet resource or technical report where a more extensive description can be found. The following is taken from Rees and Harris (1995):

> A desk-top simulator system was constructed specifically for the task, using Microsoft Flight Simulator (v. 4.0) as the driving software. The image was presented on a full color screen, giving a field of view of 45° in the horizontal and 28° in the vertical. Both instructor and volunteer subject were equipped with a full-size set of primary flight controls (stick and rudder) and a common, centre throttle. The sticks were constructed to allow operation in either a linked condition (in which the instructor›s control deflections were mirrored in the control column of the trainee) or in an unlinked condition (in which the sticks moved independently). In the latter condition, the instructor›s control column was equipped with an override facility. Following the control arrangement philosophy adopted by the Airbus A320, in both conditions the rudders remained interconnected. Flap and brake actuation was contained in a separate control panel. The flight dynamic model used mimicked the flying characteristics of a Beagle Pup 121 piston-engined light aircraft. Further details of the simulator system can be found in Harris and Rees (1993). (pp. 294–295)

Questionnaire Survey Instruments and Scales

The concise description of bespoke questionnaires can be particularly challenging to achieve. There is no definitive way of doing this. Describing survey instruments with brevity can require a degree of innovation and creativity. In the exemplar papers included in Appendices 1 and 2 the same basic strategy is used. In these instances the Method section merely contains an overview of the nature of the questions asked in the questionnaire and a description of the method for item scaling. The items themselves are described in the tables in the Results section (see table 3 in Harris and Maxwell, 2001 and table 1 in Harris, Chan-Pensley and McGarry, 2005). Remember to make sure that you call out in the text specifically to the table where the survey items can be found. This approach also avoids the use of statistical package variable names in tables (see bugbears described in the previous chapter). Unless there are compelling reasons otherwise, there is no necessity to describe other items or sections in a questionnaire that are not essential to the central message in the manuscript. Don't provide the reader with methodological details that have no bearing on the content of the Results section.

If your study uses a published, commercially available instrument such as a well-validated intelligence test or personality inventory, make sure the exact version of the scale is specified (publisher; short or long scale; revision number, etc.). Many scales may be used to collect data which are not commercially published but which are derived from the open scientific literature and are well-known. Such an instance of this is the NASA-TLX (Task Load Index) – a multidimensional scale used to measure mental workload (Hart and Staveland, 1988). A full description and illustration of the scale may not be required if the instrument is fully referenced and in common usage. However, it is those last two words that can cause problems. What is 'common usage' to one community of researchers is completely new territory for others. This is where a judgement call is required on the part of the author. Do you or don't you include a full description of the instrument? It may also be the case that the manner in which the data are subsequently treated is implicit in a simple statement in the methodology, such as *'mental workload data were collected using the NASA-TLX multidimensional scale'*. Those familiar with the NASA-TLX will know that a certain degree of computation is required to produce the individual scale scores and the overall mental workload score. Those who are not familiar with the NASA-TLX will have no idea what I am talking about. This could be exactly the same problem faced by your potential reader. If in any doubt, include all relevant detail, however it would be wise to search previous copies of your target journal to see how this issue (or ones like it) have been treated by other successful authors.

Interview Protocols

If the methodology used is an open-ended interview-administered survey where free-form interview responses are elicited from participants, give precise details of the relevant questions. Remember to include information on how and where the interview questionnaire was administered in the Procedure sub-section. On occasion, if a more complex interview protocol is used (where subsequent questions are dependent upon earlier answers) a flow chart may be considered to describe the process.

Archive Data

Although not really equipment, this seems to be as good a place as any to provide a few words of guidance on how to handle the description of archive data in the Method section of a manuscript.

Not all Human Factors research papers will involve the collection of data; some may use extant data, for example the re-analysis of accident and incident reports. A great deal of this type of data is now available on-line, especially in the aviation industry (e.g. http://www.ntsb.gov/ntsb/query.asp) but it is essential that details are provided to establish that the data used for any subsequent analysis is of good quality and from a credible source. You *must* identify the source as part of establishing a verifiable audit trail. Precise information concerning the scope of the data and the way that it was originally collected; the manner in which it is held (and retrieved); and the dates of the material in the database all help in these respects. The following is taken from Li, Harris and Yu (2008):

> The aviation accident reports were obtained from the ROC [Republic of China] Aviation Safety Council between 1999 and 2006. A total of 41 accidents and reportable incidents occurred within this time period. All accidents and serious incidents conformed to the definition within the 9th edition of the Convention on International Civil Aviation, Annex 13 (International Civil Aviation Organisation, 2006). There were 24 different types of aircraft involved in the accidents analysed, including commercial jet airliners... Full copies of all these accident reports may be found on the ROC Aviation Council web site (http:// www.asc.gov.tw/asc en/accident list 1.asp). (p. 428)

Experimental Design

In studies that use an experiment of some kind it can often be helpful to provide a concise overview of the basic experimental design, sometimes as a short, stand-alone sub-section. The following is the entire Experimental Design sub-section from Rees and Harris (1995):

> The design was an independent groups design, with linked or unlinked primary flight controls as the independent variable. Each participant was required to perform 10 simulated approaches during the course of the experiment, which formed a within-groups factor. (p. 295)

The purpose of such a description is to provide the reader with an overall framework within which to place all the subsequent details about the experimental task, the procedure and the measures taken. The description of the experimental design should also correspond to the table(s) presented within the Results section. For example, if you used a factorial design using two independent variables (one with two levels and the other with three levels) then there should be a 2 × 3 table in the Results section containing summaries of the dependent variable which matches up exactly to the experimental design described.

Experimental Task

The experimental task is a description of what you expected participants to do, the context in which they performed the trials and the relevant parameters that were applied (for example starting conditions or task completion criteria). In simple experimental studies these descriptions can be quite straightforward; however they can become quite complex and lengthy if trials are undertaken in a naturalistic setting. I have been (un)lucky enough to be involved in studies that have undertaken trials both in flight and in a flight simulator. In both cases the descriptions in the subsequent papers do not do justice to the world of pain involved in setting up and executing the trials. However, this can also be an opening to help 'sell' the uniqueness of your study. If your manuscript describes an opportunity to collect data in an unusual or rare situation, then this is the occasion to emphasise this point (and hence also accentuate the value of your contribution). You can also begin to subtly suggest how difficult data collection trials were.

As an example, the following extract from Demagalski, Harris and Gautrey (2002) describes part of the pilot's task when undertaking trials to evaluate a new emergency flight display system. This study was conducted in an engineering flight deck (a type of moderately high-fidelity flight simulator).

> For the descent sub-task the pilots were required to descend on a constant heading of 090° (inbound to a VOR). The task commenced at 11,500 feet (3,516m) AMSL (above mean sea level), 19.7nm DME (36.6km) from the beacon. The object was to descend to 8,500 feet (2,599m) on the 090° radial. To avoid exciting the phugoid pilots were advised not to use an FPA of greater than -2° (or approximately 750 feet per minute/3.8ms^{-1} vertical speed). The descent task was considered complete when the aircraft was within ± one dot (2°) on the course deviation indicator (CDI) of the 090° radial (either inbound or outbound) and was stabilised at the target altitude (±100 feet/30.6m) with oscillations contained within ±100 feet (30.6m). (p. 182)

Procedure

Finally, we get to the Procedure sub-section. This is where the recipe writing really begins. You have described all the ingredients; now tell the reader how to put them together.

The best way to deal with this section is to approach its content chronologically, from the viewpoint of either the participant or the experimenter (or perhaps a combination of both of them). From the point of view of the experimenter, what do they have to do to prepare and execute a data collection trial? From the aspect of the participant, what were they asked to do? What instructions were they given about the task and how they were required to undertake it? These instructions may extend back several weeks before the trial if, for example, they were sent joining instructions and a briefing prior to participating in the study.

Depending upon the requirements of the Journal (they differ in this respect) you may also need to include in this sub-section information about the ethical aspects of the treatment of your participants. You should provide details of any consents or waivers that participants were required to sign. The ethical standards to which the research was conducted may need to be explicitly stated:

> The research methodology was approved by an Ethics Committee acting in accordance with the guidelines for ethical conduct described by the British Psychological Society. (Ebbatson, Harris, Huddlestone and Sears, 2008, p. 1,062)

These issues will also need to be further addressed in the covering letter accompanying the manuscript submission (see Chapter 8).

Measures

From the material already written in the Results section you will have specified, either directly in the form of independent and dependent variables, or indirectly (via the sub-section on the treatment of data) all the measures that you obtained during your study. Some of these measures may already have been explained in the earlier part describing the survey instruments and scales, but many measures will not.

Some measures will be very easy to specify, for example reaction time in milliseconds; length in meters; or number of correct answers (as a percentage). However, other measures, especially those that involve scaled judgements, will be less easy to explain. These may need describing in some detail if another researcher is to replicate your work without your assistance. This short extract is taken from Dennis and Harris (1998) and is part of the description of a structured assessment scale to guide Flight Instructors when grading the performance of student pilots:

...The dimensions of trainee performance and the bounds representing the scoring parameters were derived with the assistance of the College's Chief Test Pilot and Chief Flying Instructor...

... the instructor rated the trainee on a five point scale... For exiting the turn, rolling out to within +/- 5° of the desired heading gained the maximum score of five. Exiting within +/- 10° was awarded a rating of four; +/- 15° degrees was rated as a three, and +/- 20° was awarded a rating of two. Worse performance than this was awarded the lowest rating of one. (pp. 266–267)

Note from this description not only how the scoring points were defined, but how the scaling parameters were also given credibility by calling attention to the fact that they were developed under the guidance of subject matter experts. Basically, this helps to take away any potential objections from the referees that this was not a valid measure.

Qualitative Method Description

Unlike describing the methodology for many empirical studies where emphasis is placed upon data collection, in the Method section of qualitative papers a great deal more importance is placed upon the manner in which the data are treated *after* it has been collected. Throughout the previous chapter, when the presentation and analysis of qualitative data was discussed, the need to establish credibility and confidence with the reader was consistently highlighted. This process, however, starts in the Method section.

There are two basic issues that need to be addressed in when describing the data collection methodology for a qualitative study:

- How were the data elicited, and
- Are the data valid?

Once the data have been obtained, most of the methodological concerns are related to the topics of:

- Are the data coded appropriately, and
- Are the data coded reliably?

Appendix 3 contains the paper from Huddlestone and Harris (2007) that demonstrates one way of approaching these topics.

Qualitative data are likely most likely to fall into one of two categories; verbal transcripts of some kind (either from interviews or written reports) or observational data. The reporting of a structured interview schedule has been covered in an earlier sub-section. The manner in which an interview is conducted, recorded and transcribed should be described in the sub-section on Procedure. The detailed description of the method by which interview data are recorded is essential to establish its reliability, both in terms of determining what was said when producing an accurate transcript and verifying what a participant actually meant. Data may be recorded using several methods simultaneously. For example, audio recordings may be supplemented with contemporary notes from the interviewer containing comments about the interviewee's demeanour or diagrams to aid the understanding of some technical points. You should describe the location in which the interview took place and give an indication of the typical time taken. All these details help to assure the quality of the data.

For some studies the interviewer may require particular knowledge or skills to be able to understand the interviewee's responses and on occasion, probe meaningfully for further detail. In these cases the relevant characteristics of the interviewer should also be included in the manuscript. Subject matter experts interviewing subject matter experts will help to establish the validity of the interview comments obtained.

Observational data will probably involve the categorisation or rating of behaviour, either in real time or from a video record. The key here is the precise definition of the ratings to be made or the behaviour categories employed. These should be included in the description of the methodology used.

Once the data have been obtained they may need to be categorised; they will certainly need to be interpreted. Categorisation of the data may be either driven by an existing framework (e.g. the Human Factors Analysis and Classification System – HFACS; Wiegmann, and Shappell, 2003) or the categories may be generated and developed from the data obtained (as in the paper by Huddlestone and Harris, 2007, in Appendix 3). In both cases, though, greater confidence can be placed in the Results and their interpretation if it can be demonstrated that the categorisation (or rating) process has been carried out rigorously and reliably.

The impression of reliable and valid coding of data can be enhanced by establishing the credibility of the people undertaking the process followed by describing the methods by which the raters' reliability was assessed. This extract is taken from Jarvis and Harris (2008); 'Investigation into Accident Initiation Events by Flight Phase, for Highly Inexperienced Glider Pilots':

> In accordance with previous researchers using large samples of accident data (e.g. Gaur, 2005), to establish inter-rater reliability a random sample of 100 accidents was independently categorized by the primary investigator and an independent rater. The latter was an experienced pilot of gliders and general aviation aircraft and was an airline training captain and crew resource management (CRM) instructor with training in human factors. In order to check observer consistency, the sub-sample was re-categorized by the primary investigator two weeks later to establish the intra-rater reliability (a factor omitted in many studies). (p. 217)

Note that in this instance, not only did this quotation establish the credibility of the second rater, it also emphasised a unique aspect to the study – that of establishing the intra-rater reliability in addition to the more commonly evaluated inter-rater reliability. Intra-rater reliability is often overlooked. I keep saying it (and will continue to do so): never be afraid to accentuate the parts of your paper that make it particularly distinctive. Emphasise the good points – sell it!

Li and Harris (2006) used the HFACS framework to code 523 military aircraft accidents. In this case the methodology highlighted the effort that had been put in to ensure consistent coding of the data into the HFACS framework through the training of the second rater.

> Each accident report was coded independently by two investigators: an instructor pilot and an aviation psychologist. These investigators were trained on the HFACS framework together for 10 h to ensure that they achieved a detailed and accurate understanding of the categories of the HFACS. The training process consisted of three half-day modules delivered by an aviation psychologist. The training contents included an introduction to the HFACS framework, an explanation of the definitions of the 4 different levels of HFACS, and a further detailed description of the content of the 18 HFACS categories in the context of military operations. The raters also jointly analyzed 2 yr of the ROC accident data to develop a common understanding of the process and achieve a common understanding of the categories. (p. 1,058)

In the paper from Huddlestone and Harris (2007), to build confidence in the data and subsequent model that was developed, the coding process itself was stressed. Note here, it is made explicit how the data analysis method followed the semi-prescriptive process described by Strauss and Corbin (1990) when using a grounded theory-based approach. This helped to make it difficult for any referee to argue that the prescribed methodology had not been followed. The material in tables 1 and 2 in the paper also made the process of developing the model (described in figure 2) transparent. With the use of appropriate illustrative quotes, a clear audit trail (from raw data to final model) is laid out. Again, throughout the method, a great deal of emphasis was placed of establishing the reliability of the coding process.

End of Chapter Summary

Now you have finished describing the method, if you are in any doubt about its clarity, give it to someone else to read and ask them if they could replicate it. If at any point in what follows they start a sentence with '*I don't understand this...*' what it actually means is '*you haven't explained this...*' Don't be afraid to do this.

At this point you should check that everything appearing in the Results section (people and variables) is covered in the Method. Does anything appear in the Method that isn't in the Results? If so, why is it there? Read the Method section and re-read the Results section: does the latter flow from on seamlessly the former?

It is always policy to check your 'Big Message' and list of bullet points on completing each major section. The material in the Method section should be in complete accord with the requirements of the message that you are trying to convey.

Writing the Method Section: End of Chapter Checklist

Is there a clear and comprehensive description of the target sample?	☐
Have you described all the equipment and materials used to collect data?	☐
Are all the variables in the Results section covered in the Method section?	☐
Have you removed anything in the Method section that doesn't subsequently appear in the Results section?	☐
If applicable, have you described the nature of the task participants had to undertake?	☐
Have you described the procedure employed to collect the data chronologically?	☐
Have you addressed any potential reliably and validity concerns with the data?	☐
Will your description pass the 'Delia Smith' test?	☐
Does the content of the Method section compliment the 'Big Message' and the 'story' as described in the summary bullet points?	☐
Would you want to read what you have just written?	☐

The next part in the writer's writing process is the first part in the reader's reading process!

Interlude II:
A Very Short Observation on Authorship

Depending upon where you work and who you work for, deciding upon who is credited as an author on a manuscript (and who is not) can be a very ugly debate. After that, deciding upon the order of authors can be an equally unpleasant contest. I think that I must be very fortunate in that most academics that I have had the pleasure to work with, and many whom I now count as friends, are extremely fair in this respect and always try to give credit where credit is due. However, if you are reading this book and writing the manuscript, you might want to consider putting yourself forward as being the first author (and it is always easier to crave forgiveness than to ask for permission). A few journals discourage multiple authors and set a maximum number, typically five or six. Furthermore, as a result of academic research quality audits, such as the UK's Research Excellence Framework, some academic institutions are now less enthusiastic about multi-authored papers. As a result it is worth spending just a few words here on this issue.

Few Human Factors journals have explicit criteria for authorship. However, the American Aerospace Medical Association has the following requirements for authors who wish to publish in Aviation, Space and Environmental Medicine.

> Each person designated as an author should have made substantial contributions to all three of the following: (a) conception and design of the study or analysis and interpretation of data; (b) drafting or revising the manuscript for important intellectual content; and (c) approving the version to be submitted. General supervision of the research group or participation solely in the acquisition of funding or the collection of data does not qualify a person for authorship. Those who provided technical or managerial support for the work but do not meet conditions (a), (b) and (c) should be credited in Acknowledgments.

The first author on a paper is normally considered to have made the most important contribution. This is not always the case (there may be an agreement between authors to alternate first authorship across a series of papers) but it is a reasonable assumption. In a two authored paper, it is easy to determine who comes next in the list. However, when there are more than two authors some people take this as an indication of the magnitude (or importance) of their relative contributions, in descending order. Other people are more relaxed about this. Personally, I would suggest that once you have been reduced to an 'et al.' there seems little point in fighting about your exact position in the list of credits.

In the case of multi-authored works, it is common to see in research quality audit submissions (and academic CVs associated with promotion cases) what percentage of the paper was attributable to each particular author. You might like to bear this in mind when writing the paper and agree some way of establishing it. I'm not going to even attempt to address this issue other than to comment (in my opinion) on its lunacy. Consider the overall 'percentage' contributions in the field of theoretical physics from Albert Einstein or Niels Bohr. If you look at the number of words they produced, overall it was quite small. But they were all the right words in the right order, and these *bon mots* described some pretty nifty thinking. What 'percentage' providing the fundamental theoretical impetus for the research is worth or how much value there is in solving a pivotal problem I have no idea...

I think that is enough said on this topic

Chapter 4
Writing the Introduction (Part One)

You may be relieved to know that this is going to be a short chapter as it is only concerned with writing two paragraphs: the first paragraph of the Introduction and the last one. Nevertheless, these are two very important paragraphs.

At this point you don't need your all your bullet points to hand – just your 'Big Message'. However, you also need to remember clearly what your results said (which, if you recall, was the initial stimulus for writing your paper) and the hypothesised relationship(s) that you tested. These are the crucial components around which these two paragraphs will be constructed.

Writing the rest of the Introduction will be covered later, after the approach to constructing the Discussion section has been described in the following chapter. However, the two paragraphs examined in this chapter each perform a discrete function, quite different from the rest of the material in the Introduction.

The First Paragraph

I know that it sounds like the ultimate statement of the bleedin' obvious but the first paragraph in the manuscript is probably one of the first bits that the reader will read. Remember: your initial readers will be the referees for the journal. Only after that will your work be inflicted upon the wider scientific community. Assuming that your paper has progressed from existing as a manuscript to being published, what this now actually means is that a reader has your paper in their hands (albeit metaphorical hands if they are reading an electronic version on a computer screen). It is likely that this is the end result of some sort of process which involved:

- A keyword search undertaken on some internet-based abstracting and indexing system, followed by
- Reading the paper title (and maybe checking the authors), followed by
- Reading the Abstract.

Only if the outcome of all the above fulfil some kind of criteria on the part of the 'searcher for knowledge' (they are not yet your reader) will your paper now actually be read. The other issues covered in these bullet points are considered anon, but the important thing to bear in mind is that this is now the beginning of telling the story of your research. It is your opening gambit. It needs both to create an impression and, along with the paper's title and the abstract, provide a framework for the reader's understanding of everything that follows. However, you only have relatively few words in which to do this. Many journals have a word limit for their papers, typically around 6–7,000 words. As a result, you cannot afford to waste words meeting these aims.

First of all, have a look at the 'Big Message' that you want to get across in your paper. If you read Harris and Maxwell (2001) in Appendix 1 you can see that the title of the paper is:

Some Considerations for the Development of Effective Countermeasures to Aircrew use of Alcohol While Flying

The opening lines of the first paragraph are:

The operation of an aircraft is specifically prohibited when under the influence of alcohol. No amount of alcohol whatsoever is sanctioned within the blood of on-duty aircrew. In 1985, for legal purposes, the U.S. Federal Aviation Administration (FAA) established a specific upper blood alcohol concentration (BAC) of 40mg per 100ml of blood (also expressed as 0.04%) above which it was expressly prohibited to act as a crew member of a civil aircraft. (p. 237)

From this point, three sentences in, the reader can be in little doubt that the paper is about: pilots' use of alcohol; the regulations; and approaches to deter pilots from drinking and flying. Everything that the reader then reads, particularly within the Introduction, is assimilated within this context. It provides a framework upon which to hang the information that follows. Unlike a crime thriller, where the people and the plot should slowly reveal themselves as the pages go by, a scientific paper should make it clear right from the start what it is all about. One failing I sometimes see in journal manuscripts is that the writer makes no attempt to explain in the first paragraph what their paper is concerned with. What follows is often a well written, concisely expressed analysis and synthesis of a particular area or issue in Human Factors. Unfortunately, as a naïve reader I have no idea of why I am being told this. Only a thousand words later, usually when I get to the last paragraph of the Introduction, is it made clear why I was told what I was told. I then need to go back and re-read it all to see if it makes sense once I have put it into this context. As a referee, my first recommendation to the author shortly follows!

The other function of the first paragraph is to grab your audience's attention and convince them that what follows is breathtakingly important and fantastically interesting. Ideally, the opening paragraph should be all about sex; drugs; rock and roll; smoking holes and huge body counts (the science can wait until the second paragraph). Unfortunately, such an opening to a paper is often not in keeping with the rest of the content of the manuscript. Nonetheless, this should not stop you trying to illustrate the importance of the work with reference to a significant real-world issue or event, if applicable. The following is taken from the second half of the opening paragraph in Demagalski, Harris and Gautrey (2002):

> ...Subsequently all hydraulic pressure was lost and the aircraft became almost uncontrollable in roll, pitch and yaw using the conventional flight controls. However, the crew managed to recover some control by the use of differential and symmetrical thrust changes on the remaining two serviceable engines. By using this technique the crew were almost able to effect a successful emergency landing at Sioux City airport. Unfortunately, just before touchdown the aircraft developed a very high sink rate and a slight roll. The aircraft broke up as it hit the runway with a sink rate of 1,620 feet per minute. One hundred and eleven passengers and crew were killed however 185 people survived as a result of the skill and initiative of the pilots. (pp. 173–174)

Not only does the above extract satisfy the requirement for drama, heroic actions, and death and destruction, it also introduces the reader to the background of the work in the paper, namely aircraft control using differential thrust. It is not a sin to try and make scientific writing interesting and engaging (looking at some of the papers I have written, I am beginning to regret writing this sentence...). By the way, can anyone remember the opening paragraph to this book?

The Last Paragraph

The last paragraph should be a clear statement of the aims and objectives of the research that is about to follow in the Method section. If you have done a good job as an author, the requirement for the study should be self-evident from the identification of the research need and the preceding analysis and synthesis of earlier work in the area.

For survey-based research and qualitative studies, it is often difficult to define formally a hypothesis in the same way that you can for an experimental study. In such cases I prefer to use the terms 'aims and objectives'. The following extract is taken from the final paragraph of the Introduction in Li and Harris (2006), a paper which made use of the retrospective analysis of aircraft accident reports:

> The objective was to provide probabilities for the co-occurrence of categories across adjacent levels of the HFACS to establish how factors in the upper (organizational) levels in the framework affect categories in lower (operational) levels. (p. 1057)

In Harris, Chan-Pensley and McGarry (2005) – the three-part study included in Appendix 2 – three aims and objectives (one for each section) were stated at the end of the Introduction:

> Phase One was concerned with eliciting the basic dimensions of vehicle ride, feel, performance and handling as described by a large sample of car drivers. Phase Two essentially replicated Phase One in a further independent sample to ensure that the descriptions previously elicited were stable (i.e. the dimensions had construct validity). Phase Three transformed dimensions elicited into a rating scale to assess the dynamic qualities of road vehicles and evaluated the scale's sensitivity and discriminant validity (i.e. its ability to discriminate between categories and types of vehicle in a meaningful manner). (pp. 967–968)

Outlining the various objectives for each phase in this manner also served to signpost the structure of the remainder of the paper to the reader.

Rees and Harris (1995), as it was a paper that used an experimental approach, finished the Introduction with a paragraph that summarised what the research was about and formally stated the nature of the hypothesis being tested in the Results:

> This study examines the effect of instruction using either linked or unlinked primary flight controls on the ability of *ab* initio pilots to acquire the skills necessary to perform an approach to landing. It was hypothesized that performance would he superior in the linked control condition, analogous to the control arrangement found in traditional primary flight training aircraft. (p. 294)

As a slight side issue, some US journals are now introducing a subtle change in the way that hypotheses are expressed. I only noticed this when I received final galley proofs for a paper and observed that the phrase 'it was hypothesised' had been changed to 'we hypothesised'. I found this a bit odd. I was taught that scientific writing demands the use of the neutral, impersonal, third person to emphasise the impartial position of the good scientist; but this one sentence had been changed by the desk editor and it jarred with me. I enquired why. It was explained to me that hypotheses didn't come out of thin air. They were generated by human beings (scientists) and therefore it was better to use 'we hypothesised' as it made it clear that this reflected the opinion of the authors. Obviously, the preceding 20 paragraphs of the Introduction, reviewing selected papers from a particular personal perspective but all duly written in the third person, did not reflect our opinion in the slightest!

Any hypotheses formally stated in the Introduction should directly correspond to the layout of the tables and statistical tests in Results section. Dependent and independent variables eluded to should be described in the Method section. The same basic principle should be applied to survey-based work, however with a plethora of variables it is often more difficult to enforce such strict consistency in structure across the following sections of the manuscript. Looking ahead slightly, the content of the final paragraph in the Introduction should also be reflected in the

Conclusions to your manuscript, where you pick up the success (naturally) in fulfilling the aims and objectives set out.

This just leaves one question in this section: which came first, the chicken or the egg? You may find it odd that the formal statement of hypotheses (or aims and objectives) only actually appears in the fourth chapter of the book, after both the Results and the Method section have been written. This is simply to make sure that the hypotheses (or aims and objectives) of the work described in the manuscript correspond *exactly* to the analyses in the Results section. If you remember, the central requirement in the first part of the paper writing method described in this book was 'find some interesting results'. Sometimes, for the sake of producing a nice, tight, coherent paper, a little 'reverse engineering' of the aims and objectives of a study is required.

End of Chapter Summary

Writing the remainder of the material in the Introduction will be described in a short while. However, looking ahead this is simply just a case of linking up the first paragraph, introducing the context and the 'Big Message' in your paper ('what it's all about' and 'why you did it') with the last paragraph (the introduction to 'how you did it' and 'what you thought might happen').

Once again, at this point you should check what you have written against your 'Big Message' and list of bullet points to make sure that they are complimentary. Just like a golf ball, you are 'teeing up' your reader in the first paragraph for everything else that follows in the paper. Similarly, the last paragraph containing the hypotheses, or aims and objectives, 'tees up' the reader for the success(es) that you are about to announce in the Conclusions to the paper. I can almost hear the wails of protest and the gnashing of teeth already, but writing a journal paper is not an exercise in the neutral, unbiased reporting of science. It is about putting across *your* interpretation of *your* research findings in a supportable and credible manner.

And always remember: first impressions count.

Writing the Introduction (Part One): End of Chapter Checklist

Does the first paragraph make in clear what the paper is all about; does it set out the context for the work?	☐
Does the first paragraph grab the interest of the reader and make them want to read more?	☐
Do the aims and objectives (or hypotheses) in the last paragraph reflect the analyses performed in the Results?	☐
Do the aims and objectives (or hypotheses) introduce the material that is about to follow in the Method section?	☐
Does the content of your first and last paragraphs support the 'Big Message'?	☐
Would you want to read what you have just written?	☐

The next chapter is what the science in the manuscript is all about – the Discussion.

Chapter 5
Writing the Discussion Section

The Discussion section is what it is all about. It should be the focal point of your manuscript. The whole reason for doing science is to find things out and posit a credible explanation for them. This is exactly what the Discussion section should do. What did you find? Why did it happen? In this chapter I also include writing the Conclusions and Recommendations sections. These may be either sub-sections of the Discussion or a separate major section of the manuscript, depending upon the format and common practice of the Journal for which you are preparing your paper.

A criticism I could make of many manuscripts is that they are somewhat 'unbalanced'. In the Introduction they start with a long and detailed description of the previous research in the area (much of which is of marginal relevance) but when it comes to the Discussion of the results they are somewhat cursory and excessively self-critical. Often this section is only half the length (or even less) of the Introduction despite it being the most important section of the paper. While quantity and quality are two very different (and unrelated) things, the emphasis in many manuscripts is often biased toward what has gone before rather than focussing on what the author(s) have actually discovered as a result of their labours. I find that writing the Discussion before the majority of the Introduction is one strategy that helps to redress this imbalance between the sections. The Discussion section becomes longer, as it is oriented towards interpretation and explanation, and the Introduction becomes shorter, as it can be written with a much better focus.

I also get the impression that on many occasions the authors just simply run out of steam when writing the Discussion. The end is in sight, the word limit is beckoning and so the Discussion and Conclusions get unnecessarily truncated. The Discussion section is your *BIG* opportunity to sell the quality and the importance of the work that you have done. Make it the crowning glory: give it the prominence that it deserves.

At this point I am assuming that you are familiar with the relevant Human Factors concepts and literature underpinning your work. Most journal papers are written after producing a much longer technical report or thesis. The rationale for writing the Discussion section before the remaining aspects of the Introduction is quite simple. It is to make sure that the Introduction is tightly focussed on the Discussion of the results and their interpretation. The Introduction should not be a long, rambling treatise about some aspect of Human Factors. By writing the Discussion first you will know precisely what issues to address at the beginning of the manuscript and which literature you will need to introduce to the reader to help interpret the Results and present your work in the best possible light. There is an intimate relationship between the Discussion and the Introduction. As I suggested at the end of the previous chapter, you need to 'tee up' these relevant issues at the beginning of the manuscript.

The interpretation of the 'interesting' result(s) that you found and around which you started to construct your paper is the nub of the story in the manuscript. This is the very essence of the 'Big Message'. Have a long, hard look again at the 'Big Message' that you want to convey in your paper and the outline of the story described in the bullet points. These things are going to drive the interpretation and the tone of the explanation that you provide in the Discussion. Many of the issues raised in the bullet points are also likely to appear in some fashion in this section.

There are two principle aspects to any Discussion section. Interpreting your results within the framework of your study (basically, addressing the internal validity of the study) and placing your work within the context of the wider scientific literature (examining the external validity of the work). In a great deal of Human Factors research, there is also a third aspect, which is looking at the practical significance of your results within the application domain. However, before launching into a consideration of these components, there is a very simple thing to be done first.

Remember that Your Readers are Human

And this includes your referees. By the time any reader starts the Discussion section it will be *at the very least* 3,000 words, half an hour, two 'phone calls and a cup of coffee since commencing with the Introduction. The Introduction looked at the practical and theoretical background of the work; the reader then turned their attention to assessing the technicalities of the methodology and the appropriateness/interpretation of the analysis of the data. Now their attention has to switch back to placing the results into the scientific and practical context described in the Introduction. No reader has unbounded attentional capacity and a perfect memory. Help them out a bit – give them a big headline. This will lend a hand in reminding them what your study is all about and reinforcing the scientific framework that you established in the Introduction. It will also help to get your fundamental (big) message across. For example, returning to Chapter 1, the 'Big Message' in Harris and Maxwell (2001) was stated to be:

> You need to use the right countermeasures to reduce the likelihood of drinking and flying otherwise all your efforts will go to waste – different people drink and fly for different reasons.

The Discussion section for this paper (see Appendix 1) commenced with:

> The results in Table 2 support the previous findings of Widders and Harris (1997) who observed that approximately 50% of U.K. pilots fell into either the non-believer or inadvertent drink-flyer offender categories and that holders of an ATPL were over represented in the non-believers category of potential offenders. (p. 248)

This sentence says that 50% of pilots drink and fly (i.e. this is a big issue) and that there were at least two reasons for transgressing the regulation. Half the basis for the 'Big Message' in the paper is laid down (once again) in one sentence.

Right at the beginning of the Discussion tell your reader that your study was a success and then go on to justify this comment. *Never, ever* present the evidence from your research and then allow your reader to make their own mind up: tell them what their opinion should be! The Discussion section in Appendix 4, from Demagalski, Harris and Gautrey (2002), which was a paper concerned with exercising emergency aircraft control using differential thrust under the guidance of a new flight display developed by the authors, commenced with:

> Results from the initial trials undertaken to evaluate the emergency flight control display system for flying an aircraft using only its throttles are extremely encouraging. They strongly suggest that pilots using the new system would be likely to maintain control of a severely disabled aircraft. The system would also give them a reasonable chance of executing a technically-survivable landing within the environs of an airfield that may result in only relatively minor injuries to the passengers and crew. (Demagalski, Harris and Gautrey, 2002, pp. 187–188)

So, now you have set out your stall, all you need to do is justify the statements that you have so boldly made at the beginning of the Discussion.

Interpreting the Results within the Context of Your Own Study

I'm going to open this sub-section with a couple of 'don'ts'. *Don't* simply describe the content of the preceding Results section in words ('... *the best performance was by participants in Group A, followed by those in Group C; subjects in Group B performed the worst...* '). This is not 'discussion' – it is repetition. This is a common mistake I see in many theses and subsequently in early drafts of manuscripts. The emphasis in the Discussion section must be on interpretation. The other common mistake is the converse of the previous mistake. *Don't* discuss every aspect of the results, no matter how small or insignificant. A cell-by-cell, mean-by-mean discussion of every entry in every table in the Results is not what is required.

At this point I'm going to get a little bit 'Zen Buddhist' in my approach to describing the content of this aspect of the Discussion. You can almost characterise the data and/or each analysis in the Results section as some sort of account of every individual tree and sapling in a forest. However, when you get to the Discussion section, the objective is not to describe and understand each tree but to provide a wider interpretation of the woodland vista. In general, why are the birch trees to the East of the forest? Perhaps it is because this is where the peaty, slightly acidic soils are that they favour. This is not to say that there are no birch trees over to the West (where the oaks are) but *in general* most of the birch trees are to the East in their more favourable growing conditions. No maples grow near the oak trees, as they cannot be happy in their shade.[1]

If you consider table 4 from Harris and Maxwell (2001) you can see that it is not a perfectly clean Principal Components Analysis (PCA) solution: there are some slightly aberrant loadings. However, the interpretation of these results in the Discussion concentrates on the general picture. The slightly anomalous aspects are (a) not pointed out for the benefit of the highly critical reader and hence (b) not subject to unnecessary debate.

Similarly, the Results in figure 3 in Demagalski, Harris and Gautrey (2002) did not show that the emergency flight display system worked every time, but they did show a clear trend toward superior performance. Every landing was not analysed. Note also that the figure was interpreted on behalf of the reader in a positive light with regard to the performance of the display system developed by the authors:

> ...when using the emergency flight control display system only one 'landing' exceeded a 2,000 feet per minute rate of descent. All other landings had a rate of descent of less than 1,500 feet per minute at touchdown and half the landings had a rate of descent of less than 800 feet per minute, suggesting that in these cases the undercarriage would have been unlikely to collapse. Without the display system half of the landings had a rate of descent in excess of 1,500 feet per minute, making death or serious injury a likely outcome for all on board. (p. 187)

Do not 'over interpret' your results and to reiterate, as a 'rule of thumb', never let the reader make their own mind up!

Whenever drawing any inference from your results you must always explicitly refer to the table and/or figure in which the relevant data can be found. For example:

> The broad agreement of the scale ratings with the opinions of professional motor-magazine road testers does suggest that the scale aids in producing valid ratings of the dynamic behaviour of motor vehicles (**see tables 8 and 9 and figure 2**). (Harris, Chan-Pensley and McGarry, 2005, p. 980 [emphasis added])

In doing this you are establishing a verifiable audit trail to support the assertions that you are making in the explanation of the results. Also note in the above extract the subtle use of slightly loaded language to 'help' the reader interpret the results in the

---

1   Apologies to Geddy Lee, Alex Lifeson and Neil Peart.

same way that the author did (i.e. me). You always need to 'sell' your paper a little bit: tell the reader how good your results are.

Whenever considering a finding in your data along the lines of *'people in condition A, on average, performed significantly faster than those in condition B'* you should always proffer a reason for this, otherwise you are falling into the trap of description and not discussion. *Why* did they perform better? For example, in Demagalski, Harris and Gautrey (2002) we showed that pilots performed better as a result of using the emergency flight display we developed. However, the discussion in the paper went further than just this and suggested a potential mechanism by which this happened:

> ... the new display system had a predictor element incorporated into it in terms of the commanded throttle position required to achieve a certain altitude or heading, this overcame many of the control lag problems that induce poor performance. (p. 189)

The mark of a good Discussion is that it provides *explanation*.

Interpreting the Results within the Context of Previous Work

Interpreting the reasons for obtaining your results is only half the battle. You also need to place your work within the wider context of the findings from previous authors. Not only does this help to give your work credibility it also helps to develop the wider science base and may begin to establish a common underlying principle. The generalisability of scientific and theoretical concepts is important and this issue should be touched upon in the Discussion.

When writing the Discussion section my personal approach is to interpret the Results in the context of the study and write these aspects first. I then go back to the literature and insert in (as required) references to other authors' work, weaving it into the narrative that I am developing about the study I am describing. By doing this I hope that it helps to keep the work being presented in the paper as the central focus (the 'spine' of the Discussion section) and not the work of the earlier authors. Remember: the majority of the content of the Discussion section should be about your own work: it's your showcase.

When the results from your study complement the work of previous researchers, say so. Simply call out to the work of these authors (cite them). The findings from these people should already have been described in the Introduction to the manuscript (but you haven't written this yet...). There is no requirement to describe it all again. However, do bring to the attention of the reader the underlying commonalities in the work of others and your own study. To illustrate, the following excerpts are taken from Harris and Maxwell (2001) where explicit parallels were drawn with the findings from research published by earlier authors:

> The PCA performed on the opinions about the potential effectiveness of the various countermeasures produced four factors each relating to a different aspect of the tripartite approach described by Vingilis and Salutin (1980): at the primary level, the education of pilots concerning the effects of alcohol; at the secondary level, strategies for the enforcement of drinking and flying regulations; and at the tertiary level two factors were produced, one concerning the counseling and rehabilitation of pilots with personal problems and a second factor concerning the severity of sanctions for offenders (see Table 4). (p. 249)

> ... sanctions were more likely to be regarded as an effective deterrent by the section of the flying population that was less likely to offend. The fact that sanctions were regarded as a less effective countermeasure by both offender groups compliments the findings from the drink-driving literature (e.g., Mäkinen, 1988; Ross, 1988; Wheeler & Hissong, 1988; McKnight & Voas, 1991). (p. 249)

The key thing that made this particular study new was that the work was carried out in a different application domain, i.e. the deterrence of pilots drinking alcohol and flying rather than drivers' drinking and driving. This was a large part of the paper's scientific contribution. This should come across very clearly (again) in the Discussion.

If the results from your work builds upon previous studies (especially your own), taking it a step further, then say so.

> Maxwell and Harris (1999), also found that possessing a professional pilot's license was a strong determinant of drink-flying offending. In this later survey, again over 50% of respondents reported that they had (or may have) flown when their BAC was in excess of the prescribed 0.02% limit giving further credibility to these results. (Harris and Maxwell, 2001, p. 248)

Another common outcome is that your work contradicts the findings from earlier work (which you will have described in the

Introduction). It is not a disaster if your results disagree with these authors: in fact quite the reverse. It can provide a great deal of material for discussion and for developing the general understanding of the underlying phenomena at work. The manner(s) in which your study and the other studies differ are probably key variables for further exploration. In such a case, cite the contradictory studies and simply outline the nature of the discrepancy in results between the studies, but again and most importantly, suggest a reason for the difference. Do not denigrate the earlier work, though, suggesting that it was wrong or misinformed. This will win you no friends. Simply point out what you feel are the relevant differences that may explain your findings. Again, this can be illustrated by drawing upon the paper by Harris and Maxwell (2001) as an example:

> When comparing the results of this study and earlier surveys of U.K. pilots (Widders & Harris, 1997; Maxwell & Harris, 1999), to similar studies of U.S. pilots (e.g., Ross & Ross, 1992; Ross & Ross, 1995), there would seem to be major differences in the pattern of results. The U.K. studies suggest a higher rate of offending and different opinions concerning what would be the most effective countermeasures. It is not clear if these findings relate to cultural differences between the two populations or are responses to the different regulations. It can be speculated that the higher level of offending reported in the U.K. is a product of the much lower BAC prescribed in the revision to JAR OPS. Ross and Ross (1995) commented that in the U.S., the 0.04% rule has been criticized as being too lenient. It is interesting to speculate if the same pattern of results as those found in this study would be observed in U.S. aviators if the upper BAC limit was reduced to 0.02%, as in Europe. (p. 250)

Note in the preceding paragraph, just like in the Introduction, the emphasis is very much on analysis and synthesis and not just simply on description.

The Discussion of Results from Qualitative Analyses

The discussion of the results of qualitative analyses often proceeds in a different manner from papers employing the statistical analysis of data. To an extent, this can depend upon the requirements of the journal for which you are preparing the paper (which is one reason why you should familiarise yourself with your chosen publication). Most journals still require a Discussion section of some kind, however in many cases the analysis and interpretation of the data are undertaken side-by-side throughout the Results section (see Chapter 2). The analysis and interpretation of the data may refer to works cited earlier in the Introduction in order to provide explanatory power. This is particularly the case if using a methodological approach based upon Grounded Theory (Glaser and Strauss, 1967; Strauss and Corbin, 1990). However, the purist 'Grounded Theorist' is completely driven by the process of analysing the data and approaches the whole analysis from a 'theoretically naïve' position. After the data have been analysed only then are parallels sought from the science base to aid interpretation. In practice, this hardly ever happens as it is almost impossible to approach such a process in such a naïve manner.

However, the final Discussion should still concentrate upon placing the analysis of the data categories obtained and the model developed within a wider practical and theoretical context. Perhaps slightly less emphasis is placed upon interpretation in the discussion of qualitative data (compared to quantitative data) as this has already been undertaken in the Results section. In this case the accent in the Discussion section is more concerned with placing the work within a wider scientific framework. For example, in Huddlestone and Harris (2007) the model developed is validated in the Discussion section with explicit reference to parallels found in the work of earlier researchers:

> Figure 2 shows the complete performance model, with situation awareness, decision-making and action categories broken down into their sub-components. The central feature of the model is the constant repetition of development of 'situation awareness' leading to 'decision-making' and subsequent 'actions'. 'Communication' is prominent, both as an aspect of decision-making and as a required action following decision-making. These features of the model were consistent with the findings of Waag and Houk (1994) in their development of a set of behavioural indicators suitable for evaluating proficiency in air combat in day-to-day F-15 squadron training. They identified communication, information interpretation and decision-making as key activities. This point is reinforced by Bell and Lyon (2000) who reported that communication was one of the most highly rated elements contributory to good situation awareness, based on a survey of mission ready F-15C fighter pilots. The monitor-evaluate-anticipate model of situation awareness was found to be directly equivalent to the perception-comprehension-projection model of situation awareness proposed by Endsley (1995). (pp. 365–366)

Nevertheless, the main objective of the Discussion section for qualitative data is the same as for any paper. It should provide explanation for the findings.

In other cases, where the analysis of qualitative data is driven by some kind of model (as in Psymouli, Harris and Irving, 2005) the first aspect of the analysis/discussion is to legitimise the analytical framework (as noted in Chapter 2). Having done this the categorisation process should refer explicitly back to the model. The inferences drawn in the analysis/discussion section is usually simply concerned with interpreting the data within the context of your own study (discussion of the findings can wait until the 'real' Discussion section). To illustrate, Psymouli, Harris and Irving (2005) is a paper concerned with inspection of aircraft composite structures during scheduled aircraft maintenance inspections. As the title of the paper suggests, the interpretation of the results is undertaken using a Signal Detection Theory-based framework. The following observation was made from one participant contributing to the study:

> '…if the inspection needs to be conducted during a particularly windy evening, I will have to place my cherry picker at a greater than the normal distance in order to avoid an impact of this with the aircraft, which will be moving due to the wind. However from such a distance I might not be able to detect all the existing defects. […] if the sun is shining very brightly into my eyes and I am trying to inspect the rudder I might miss something during that particular inspection'.

This was then interpreted within the theoretical stance of the paper, thus:

> Effectively, in signal detection terms, what is happening in these instances is that by being further away the signal strength (i.e. damage) is being weakened (moved to the left on the X axis in figure 1) by increasing the inspection distance. This means that if the decision criterion remains 'fixed' more signals are likely to be missed or if the decision criterion moves to the left to compensate, there will be a commensurate increase in the false alarm rate. (Psymouli, Harris and Irving, 2005, pp. 97–98)

As in the earlier qualitative paper referred to (Huddlestone and Harris, 2007) the Discussion section then proceeds to place the results obtained within a practical and theoretical context.

A Major 'Don't'

At the end of the Discussion section it is notable how some authors seem to take great delight in shooting themselves repeatedly in the foot. They load, take aim and fire a volley of unnecessary self-criticism at the work that they have just completed. This can either serve to draw the referees' attention to methodological shortcomings in the manuscript that they had not previously noticed, or simply leave the reader with a lasting poor impression of the work. First impressions count but so do the last impressions that you leave the reader with: always try and leave them on a high. You have one last chance to do this in the following Conclusions and Recommendations section but don't start creating doubts about the quality of your work at the end of the Discussion.

There is a fundamental difference between a thesis and a scientific journal paper. In a thesis, the examiner is looking for evidence that the candidate is aware of any potential methodological limitations and/or shortcomings. In short, the examiner is checking that they are a good scientist. Referees of journal manuscripts are, amongst other things, checking to see if the work is of acceptable scientific quality to be placed in the open scientific literature. If you are aware of any fundamental flaws in the work you are trying to publish, then why are you writing the manuscript? However, most scientific works have minor points that could be improved upon. In some cases you may wish to open this debate concerning the methodological choices made (and hence put your side of the argument first) as a way of taking away any ammunition from your critics. However, consider doing this in the Method section, perhaps as a sub-section on Methodological Considerations.

In general, don't advertise minor methodological or analytical shortcomings to the referees, though. If the referees happen to notice some issues in the methods or analyses that you have employed, then address these points in the revised version of the paper (see Chapter 10). Don't open the debate unnecessarily at Round One. Always remember to accentuate the positive aspects of your paper and don't draw attention to the more debatable aspects of your work. As I keep saying throughout this book, you need to 'sell' your work. Getting your paper published is as much a problem of presenting your work in a positive light as it is about the scientific quality of the material in it.

Conclusions and Recommendations

The Conclusions mark the place where many authors get tired of thinking. Don't fall into this trap. In a good paper the conclusions to the study should relate directly to the aims and objectives set out at the end of the Introduction. They should be short, sharp and to the point. It is basically your chance to say in no uncertain terms that the study was of value. The Conclusions should also relate to the 'Big Message' underlying the whole of the paper. Earlier in this chapter I provided a reminder of the 'Big Message' in Harris and Maxwell (2001), so to avoid repetition I'm not going to do it again. In this paper the 'Big Message' was then translated into the following objective for the study:

> This study examines the relative effectiveness of the components of the tripartite model for the deterrence of drinking and flying with respect to the type of offender and license category of pilot. (Harris and Maxwell, 2001, p. 240)

The opening of the first paragraph of the Conclusions section starts:

> As Ross and Ross (1995) noted, it is important to take account of pilots' opinions about what constitutes effective countermeasures as they have the most experience of the use of alcohol before flying and best know what would be most likely to deter them. In the U.K., the results from this survey indicate that efforts would be best placed in developing enforcement and sanctions countermeasures (the secondary and tertiary aspects of the tripartite model). These should primarily be aimed at holders of an ATPL who are over represented in the non-believers offender group. (pp. 250–251)

Put another way, the whole of the manuscript comes 'full circle' to where it started off in the Introduction: it is all internally consistent in communicating the desired message to the reader and presenting the work in a positive light.

The Recommendations component to the paper may have two aspects to it: practical recommendations, following on from the findings of the study, and scientific recommendations concerning how to follow up the work in further research. Spell both types of recommendation out clearly for the reader. These will also probably directly relate to the impetus for undertaking the work in the first place and may well reflect the material in the opening paragraph of the Introduction, which set out the context for the study. To illustrate, if you refer to the Huddlestone and Harris (2007) paper contained in Appendix 3 you will see that its title is:

Using Grounded Theory techniques to develop models of aviation student performance

The opening paragraph of the Introduction commenced:

A frequently encountered issue in aviation training research is that of student performance modelling. A potentially rich source of student performance data exists in many if not all aviation training organisations in the form of narrative reports written after flying or simulator sorties. The challenge is in finding a suitable way to sample and then analyse the data in these reports to produce a meaningful model. (p. 357)

The first conclusion and recommendation in the paper was:

...the grounded theory analysis clearly identified eight basic areas (and nine further subareas) for meaningful assessment of student performance in the Tornado F3 pairs phase. Therefore, the model provided a sound basis upon which to determine candidate measures for a subsequent training effectiveness trial. (pp. 367–368)

With regard to making scientific recommendations about follow-on research, this can be your opportunity to introduce the next paper in the series that you are going to write (even if the work may already have been done). It also helps to establish your claim to the idea. To this end Harris, Chan-Pensley and McGarry (2005) concluded:

Further development of the scale is planned in a series of trials in an engineering simulator and in a set of test-track trials. These trials will further evaluate the criterion validity of the scale and also assess its test – re-test reliability and sensitivity. (p. 981)

End of Chapter Summary

Discussion is about providing an interpretation of your results and an explanation for your findings. It is what science is all about. The Discussion should place your work into the practical and scientific context that (will be) laid out in the Introduction (when you write it). You can almost think of the Introduction and Discussion sections as bookends, but bookends *on the same shelf.* They bracket the work in the paper and they should be completely complimentary.

Don't just let your manuscript fizzle out at the end: finish on a high note with strong conclusions and set out a positive direction for future research and applications. I'll say it again: this is the last thing that the referees will read before starting their final assessment of your work. Create a good, final impression. Sell the content of your paper and convey the idea that you are a good scientist with lots of ideas to develop the work further. Give the reader confidence in you.

Writing the Discussion: End of Chapter Checklist

Do you open the Discussion section with a gentle reminder concerning what the paper is all about?	☐
Have you interpreted the meaning of your results and explained why you think they happened?	☐
Do your findings compliment earlier studies by other authors (if 'yes', then say so).	☐
Do your findings extend earlier work, especially your own (if 'yes', then say so).	☐
Do your findings contradict earlier work (if 'yes', then say so and suggest a reason for this).	☐
Have you finished off with a strong conclusion that relates directly to the aims and objectives?	☐
Have you made any practical recommendations and/ or suggestions for further work?	☐
Do you end your paper on a high note?	☐
Does the content of the Discussion and Conclusions sections compliment the 'Big Message' and the 'story' as described in your summary bullet points?	☐
Have you removed anything in the Discussion and Conclusions sections that is irrelevant to the 'Big Message' and the 'story' as described in your summary bullet points?	☐
Would you want to read what you have just written?	☐

The next chapter will be all about introducing the concepts at the beginning of the paper that you have just been referring back to in the Discussion. But you haven't written this yet. I'm beginning understand what Douglas Adams was talking about.[2]

---

2 'One of the major problems encountered in time travel... is simply one of grammar... the main work to consult in this matter is Dr. Dan Streetmentioner's *Time Traveller's Handbook of 1001 Tense Formations*. It will tell you, for instance, how to describe something that was about to happen to you in the past before you avoided it by time-jumping forward two days in order to avoid it. The event will be described differently according to whether you are talking about it from the standpoint of your own natural time, from a time in the further future, or a time in the further past and is further complicated by the possibility of conducting conversations while you are actually travelling from one time to another...' (Douglas Adams, *The Restaurant at the End of the Universe*).

Chapter 6
Writing the Introduction (Part Two)

You have already written the first paragraph of the Introduction (introducing the context and the 'Big Message' in your paper) and the last paragraph of the Introduction (the aims and objectives). Now all you need to do is write a few relevant paragraphs linking the two together.

In the preceding section I mentioned that there is often a considerable 'imbalance' between the Introduction and the Discussion, with far too much emphasis placed on the former and not enough on the latter. Let's just look at that issue a little further. In my role as an internal or external examiner on Human Factors courses across the UK I have come across several marking guidelines for theses, and all of them award far higher marks for the Results and Discussion sections than they do for the Introduction. It is not uncommon for only 10% of the total thesis mark to be available for work in the Introduction but 50% of the marks to be available for the Results and Discussion (combined). There is a *big* difference between writing a thesis and a journal paper but it is still worth bearing this in mind when preparing your manuscript. It gives you a good idea of the notional 'value' of the various sections.

In the comments to a paper that I submitted several years ago I was once asked (by the Editor of a medical journal) why a particular citation appeared in the Introduction that was not subsequently referred to in the Discussion? What merit was there in citing this author's work? Could it be deleted to shorten the manuscript? While not all journals are quite so fastidious in this respect, it is a good question that you should be asking yourself when it comes to writing the Introduction. If a reference appears in the Introduction that is not subsequently referred to in the Discussion, then why is it there? What is it adding to the manuscript? Similarly, if material appears in the Discussion that is not initially referred to in the Introduction, then why not? Surely this material must be of scientific or contextual importance in providing an explanation for some of the Results? In which case, it should appear somewhere at the beginning of the manuscript.

As is usual at this point, it is worth re-visiting your 'Big Message' and the story of your paper as described in your series of bullet points. Go and re-read your Discussion section as well. Having done that, the time has come to start writing again.

Material that may Need to be Covered in the Introduction

Before going any further, at this point it may be useful to make a quick inventory of the material that you might need to include in your introduction. The purpose of the Introduction is to set out the theoretical and applied context for your Human Factors study. However, there are several aspects to achieving these objectives. These include:

- A description of the applied/practical background to the problem (for example in the form of statistics or case studies).
- An analytical consideration of the theoretical background to the problem (as described in the previously published literature and particularly the literature referred to in the Discussion section: make a list of these references to help you when writing the Introduction).
- A justification (either explicit or implicit) of the methodology that will be described in the Method section.
- A justification (either explicit or implicit) of the dependent and independent variables used (or survey items, published inventories, etc.).
- A justification of the sample used in the study.

Do not consider these bullet points to each represent a paragraph or sub-section. They are simply issues that may need to be covered as they are all important parts of the study that need to be justified to the critical reader. However, this is not intended to be an exhaustive list. It is also worth having a quick look back at the bullet points at the opening of Chapter 3 in this book. Remember, you are trying to construct a coherent, internally consistent manuscript, not a paper comprising four or five sections all somehow relating to the same topic.

Practical Background

The practical/applied aspect to the problem should already have been introduced in the opening paragraph to the paper which sets the scene for the reader, but you might need to go a little further to develop the point. Don't be tempted to include too much detail at this point. *Anything* that you include in the Introduction must (metaphorically) pay its way. Material that is included just to add a little 'colour' and to help bring the manuscript and its contents to life must be used sparingly. This is not to say don't do it – just be careful. The next extract is taken from a paper by Bohm and Harris (2010) which looked at accidents on construction sites involving dumper trucks. It was assumed that most readers would not be aware of the dangers imposed by this type of equipment so the following was intended to provide a little background to promote their understanding of the issues:

> The following accidents are typical and illustrate some of the risks: an untrained dumper driver was killed when he was thrown from the dumper after it hit a shallow trench; a dumper driver was killed when a forward-tipping dumper overturned on a slope after reversing; another driver was killed when his dumper ran off the road and overturned in a ditch. (p. 55)

The issues implicated in accidents like these were re-visited in the construction of the stimulus material used in the study (described in the Method section) and again in the Discussion section.

A short review of appropriate statistics (for example, accident statistics) can be useful in establishing the size/importance of the issue at hand. However, these figures should be from reputable sources (e.g. governmental statistics). Did I just write that? A short analysis of statistics over a period of years may establish a trend which can bolster any argument for the necessity of your research. The numbers can also be used to surprise the reader (to some very mild extent) bringing something to their attention that they were not previously aware of, or correcting a commonly held misconception, again giving further credence to the research that follows:

> A study of 143 US gliding accidents found that 'landing' accounted for over half of all accidents but was only associated with 10% of fatalities. The 'cruise' flight phase accounted for the highest proportion of fatal accidents (36%) (van Doorn and de Voogt, 2007). (Jarvis and Harris, 2010, p. 294)

However, do not go 'over the top' in filling the Introduction to your manuscript with compelling statistics to develop the case about just how important your work is. You should always bear in mind that in a Human Factors journal, even though it may have a very strong applied remit, the emphasis throughout should be on the scientific and theoretical dimensions of your study. Use statistics judiciously.

Keep referring back to your 'Big Message' and bullet points to check that anything you have written about previous accidents or statistics is 'on message'. If something that you have written doesn't add value to the story that you are trying to tell, then delete it.

Theoretical Background

The theoretical background to the study should occupy the vast majority of the space in the Introduction. This material will set out the scientific context for the paper and hence help to provide the explanatory power for the interpretation of the Results provided in the Discussion. As a first step, assemble all the references cited in the Discussion. Each one of these *must* appear somewhere in the Introduction.

It is very tempting to try and weave the Introduction around the content derived from previous papers. However, this can be very difficult and it can result in an incoherent narrative containing many side issues and more 'exceptions' than 'rules'. My personal approach is to write the 'story' in the Introduction based simply from my understanding of the scientific area and the research problem to be addressed. I also try and do this very quickly, almost as if I was telling the story to someone. For me, this makes it flow better and it helps to keep it focussed (but I realise that this won't work for everyone). The next step is to put in references to support all the assertions made. I should also point out that I make no attempt whatsoever to provide a comprehensive and unbiased view of the previous literature. The works from earlier authors included in the Introduction are there to serve *my* purposes (i.e. providing a theoretical basis for the work that follows, and offering insights to help in the interpretation of the findings). If necessary, potentially contradictory literature is simply omitted or may be glossed over. Sometimes including it and then pointing out its methodological shortcomings, etc. (from your point of view) merely draws it to the attention of critical readers and potentially begins to beg questions of your work. Furthermore, if you can't really say something positive about previous author's work, it is often better to say nothing at all. However, presenting both sides of an argument (if this is the case) can allow you to draw strong conclusions supporting one side (or the other) in the Discussion section.

The critical tool for constructing the argument in this section is the piece of paper containing the paper's intended 'Big Message' and the bullet points that break the story down a little further. Referring back to the first three bullet points in the story underlying the Harris and Maxwell paper (described in Chapter 1) opening, these were:

- Drinking and flying is undesirable – new regulations are on the horizon (at the time of writing).
- A great deal is known from the development of drinking and driving countermeasures but these have never been applied to aviation.
- Effective countermeasures depend upon the reasons for the underlying offending behaviour.

The first bullet point was covered in the opening paragraphs of the paper:

The operation of an aircraft is specifically prohibited when under the influence of alcohol. No amount of alcohol whatsoever is sanctioned within the blood of on-duty aircrew. In 1985, for legal purposes, the U.S. Federal Aviation Administration (FAA) established a specific upper blood alcohol concentration (BAC) of 40mg per 100ml of blood... Until April 1998, Article 57 of the U.K. Air Navigation Order (No. 2; Department of Transport, 1995) merely stated that "the limit of drinking or drug taking is any extent at which the capacity to act as a crew member would be impaired." This regulation was subsequently amended to incorporate a revision to the European Joint Aviation Authorities operations regulations (JAR OPS) that specifically prohibits a pilot to act as a crew member with a BAC of greater than 0.02%. For the average 80kg man, this corresponds to drinking just over 1/2 pint (254ml) of normal strength beer. (Harris and Maxwell, 2001, pp. 237–238)

It can be seen from the second bullet point that the premise for this paper was essentially one of a 'transfer of technology' from the road to the aviation environment:

The study of drinking and driving has a great deal to offer in the development of effective measures to discourage drinking and flying, however, little of this literature has been referred to in this context. Vingilis and Salutin (1980) described a tripartite model for the deterrence of drink-driving behavior... (p. 238)

The third bullet point looked at the general methods by which it had been established in studies of drink-driving, that countermeasures could be made more effective:

The effectiveness of general deterrence as a drinking and driving countermeasure has been difficult to establish. Mäkinen (1988) reported that from a review of over 200 enforcement studies, severe punishment had no demonstrable effect on improving drivers' behavior. Ross (1988) supported this finding, suggesting that increasing the likelihood of apprehension on a drink-driving occasion (Stage 2 of the model) was a more effective deterrent. However, later research has suggested that widely advertised punitive sanctions could also act as a deterrent to initial offending (Kinkade & Leone, 1992). The implementation of severe sanctions for drink-driving in California reduced the arrest rate by approximately 12%. The effectiveness of punitive sanctions as a specific deterrent (i.e. applied to offenders to deter re-offending) would also seem to be limited. Wheeler and Hissong (1988) observed no difference on the likelihood recidivism with respect to the severity of the sanction imposed (probation, fine or imprisonment). (p. 239)

Note that throughout the Introduction to this paper (and others) there is little or no comment about the methodologies employed. Unless you are intending to make a major methodological point about the previous research (which may be associated with the methodology that you have chosen to employ) or you are writing a methodology paper, there is no requirement to give any details in this respect. The emphasis should be upon what the authors of previous studies have found that is of relevance to the research that follows. When writing the Introduction the key bits from each of the papers reviewed can usually be found in two, quite short sections. Check the papers' Abstract and the final Conclusions.

The level of detail in each of these sections is usually more than enough for writing an Introduction.

However, a common weakness in many Introductions is that they are merely descriptive. They just provide a historical series of brief synopses of earlier research. This is not what is required in a good Introduction. There should be analysis and synthesis (most particularly the latter). By grouping together studies that have common (or complementary) findings, a strong basis for the research in your manuscript can be presented. Draw out the common thread(s) underpinning their findings:

> The effect of enforcement (secondary level intervention) on suppressing drink-driving behavior is well understood. Increasing the perceived likelihood of arrest acts as an effective deterrent (Guppy, 1988). Increasing the actual likelihood of arrest when drink-driving also suppresses this behavior. The introduction of large-scale random breath testing (RBT) in Finland and Australia led to beneficial effects on the suppression of offending behavior (Dunbar, Penttila & Pikkarainen, 1987; Homel, Carseldine & Kearns, 1988). In Finland a 58% reduction in the level of drinking and driving was observed after the introduction of RBT; analysis of Australian data showed a 42% reduction in the number of alcohol related road traffic accidents. (Harris and Maxwell, 2001, p. 239)

If you want to save yourself a lot of time and effort in collating the material for your Introduction (and Discussion) and have some of the analysis and synthesis done for you (it saves all that thinking) it is a good idea to search for review papers in the domain of interest. For example, anyone who is looking to undertake research in the area of drinking and flying should look at the excellent critical review in Harris (2002: 2005)! However, do read the original papers – don't just work from the review article. The same things are interpreted in different ways by different people working in different contexts. An often quoted study on the effects of alcohol on short-term memory was actually undertaken on goldfish (not people). If you cite something, make sure that you have read it.

Whenever you review literature and draw out commonalities and conclusions from earlier work, you then need to tell the reader what implications they have for your study. Again, be quite explicit. Tell them why you are telling them all these interesting things. *Never* let the reader draw their own conclusions about what you are saying (but do try to do this subtly):

A fundamental problem with the use of this scale, however, was that the complexity of the aircraft's behaviour was not reflected in the dimensions of the scale (Payne and Harris 2000). In addition, the Cooper-Harper scale did not describe the interaction between the aircraft's dynamic qualities and the nature of the task being undertaken. What may be desirable qualities in one type of manoeuvre may not be so desirable in another type of manoeuvre (see Harris et al. 1999, 2000, Harris 2000). Similar criticisms may also be made of unidimensional scales to describe the handling qualities of motor vehicles. What may be a desirable characteristic in one type of vehicle (e.g. a more softly sprung, compliant ride in an executive car) may not be so desirable in another category of vehicle (e.g. a sports car). Unidimensional scales do not take into account the interaction between vehicle type, its dynamic road behaviour characteristics and its intended market segment. (Harris, Chan-Pensley and McGarry, 2005, p. 965)

Whenever you make any major assertion in the Introduction you will always need to back it up with citations to earlier studies. It is always a good idea to try to include some references from the journal for which you are preparing your manuscript. This helps to give the Editor and reviewers confidence that you are familiar with the remit of the journal and it also helps to improve the all-important Impact Factor (see Chapter 1). The latter will make them very happy. Don't include non-applicable references, though.

As a point of style, use quotations from other authors only sparingly (and when you do this make sure you credit them properly). Quotation *is not* explanation. As a 'rule of thumb' you should only ever use a quotation when you cannot say it better or more succinctly yourself. Furthermore, if you use too many quotations they lose their impact.

Not all citations to previous works are of the same worth. There is a hierarchy in value. At the top of the tree, naturally, are peer reviewed journals. After this (in my opinion) comes material sourced from technical reports originating from reputable laboratories (e.g. NASA) or textbooks (including edited books with individual chapters invited from prominent authors). Remember textbooks are not really peer reviewed and they usually contain second hand (at least) synopses of research. Material from textbooks is, however, very useful for providing a wider theoretical framework. Information sourced from conference papers should only be cited with care. There are conferences that have extremely high standards for the selection and publication of papers, some of which are almost equivalent to those of a top flight journal in terms of their production and peer review (the Academy of Management and the Interservice/ Industry Training, Simulation and Education Conference – I/ ITSEC – are two that spring immediately to mind). And there are conferences that don't have such lofty standards. In some cases the two most important requirements for having a paper accepted are (a) you can pay the conference fee, and (b) you can reliably remain upright while talking. Finally, there is the internet… source material from here using immense care and discretion.

It is worth bearing this hierarchy in mind. Cite the best possible sources that you can to support your argument, but you don't have to reference every source. Some journals place a limit on the number of references that a paper may include (usually in these cases there is an upper limit of 25–30). In such an instance, use only the most directly relevant and higher quality sources to support your case.

Methodological Considerations

On occasions, the methodology employed is of particular relevance to the content of the manuscript. The methodology may be novel or it may be a new application of a research approach developed in a different domain. In cases like these it is necessary to address this issue directly somewhere in the paper, as it is a central tenet underlying the work. This can be done either as part of the Introduction, or perhaps as a sub-section at the beginning of the following Method section. In either case, though, the objectives and content are the same. You need to establish that the approach adopted is legitimate for the research in question. Only if the referees of the paper are convinced of this with they regard the Results as being valid.

> With regard to the scaling of risks the simplest option is to ask respondents to rate the level of risk for each hazard on a scale [20]. This approach is intuitive and relatively easy for participants (especially with low numbers of items) but it lacks rigor. A superior alternative is the paired comparison technique (based upon an extended card-sort), which simply requires respondents to compare each item with every other item until every permutation of paired comparisons has been exhausted. This has well-founded theoretical underpinnings. It is based on Thurstone's law of comparative judgment, which contends that scaled judgments can be made for practically any attribute [24]. The advantages of this technique are that it has intrinsic rigor and it can also provide insights into problems with the scaling of items and/or the judgment of the rater…

Both Ostberg in a study of the perceived occupational risks in forestry workers [21] and Weyman and Clarke in a study of miners [22] used this approach. Participants were required to judge a series of paired, carefully selected and industry-specific risk scenarios represented in the form of line drawings plus explanatory text. (Bohm and Harris, 2010, pp. 56–57)

The previous extract (which was part of the Introduction) basically makes the case that (a) this is methodologically a good way of doing it; (b) the methodology has an established theoretical base and (c) it has been done this way before but in slightly different contexts. What more could you ask for? Well, the best bit is about to follow…

In both cases [the Ostberg and Weyman and Clarke studies] a high level of agreement was observed in the risk perception scales derived. Unfortunately, in neither study was the accuracy of the perceived risks compared with an objective measure of the same risks derived from accidents as these data sets were either unsuitable or unreliable. (Bohm and Harris, 2010, p. 57)

Here is a gap that has been observed in the previous research. Guess what the Bohm and Harris paper sets out to do? Remember: one of the main roles of the Introduction is to provide the basis for the material in the Discussion of all the noteworthy bits in your research. It should 'tee up' these issues for the reader.

Justification of the Variables Used

In a similar way to needing occasionally to justify the methodology employed, it is sometimes necessary to provide a good reason for the variables collected, especially the dependent variable(s). This is particularly the case when an alternate dependent variable is available, perhaps one more commonly used.

When evaluating performance on any tracking task, such as flying an aircraft, it is most common to examine the end product of performance (i.e. measuring errors between the tracked parameter and a target value). Metrics such as the arithmetic mean error and standard deviation of error have strong validity when applied to parameters such as flight path or airspeed deviation especially when associated with a well-prescribed flight task that demands a high level of performance, such as flying an ILS-approach. The arithmetic mean error gives an indication of the overall flight path error (on a particular axis) and its associated standard deviation gives a measure of the 'smoothness' of the pilot's performance. These two parameters are often used in preference to the Root Mean Square Error (RMSE). Taken in combination, the arithmetic mean error and the standard deviation of error completely define the root mean square error. Furthermore RMSE also has the additional disadvantage that it produces identical values for quite disparate performances. For example, being consistently high, consistently low, or at the correct mean height but with great variations in height keeping may all result in the same RMSE value (see Hubbard, 1987). (Ebbatson, Huddlestone, Harris and Sears, 2007, pp. 370–371)

When I am reviewing papers, I often find myself asking 'why has the author chosen to do it that way': 'why didn't they do something else'? Providing a critical commentary on the methodology applied and the variables used immediately removes these questions from the mind of a reviewer.

Justification of the Sample

Very often, this is not required. However, if there is an analysis based upon a contrast between sub-samples in your Results section (e.g. in Harris and Maxwell, 2001, there is a comparison between the attitudes towards drinking and flying countermeasures between private and professional pilots) it should be obvious from the material in the Introduction what the basis is for this analysis. The sub-sections in your sample become your independent variables. In other words, you need to introduce the requirements for the sample of participants that you are about to use in the following sections. For example:

Maxwell and Harris (1999) used a structural equation modelling approach to predict drink-flying offending behavior. Offending behavior was best described as a combination of personal factors (e.g., indoctrination into the culture of professional aviation) and situational factors (e.g., job-related stresses). Professional pilots holding an airline transport pilot's licence (ATPL) were found to have the highest mean weekly consumption of alcohol. Sloane and Cooper (1984) also reported that 12% of professional pilots drank as a means of coping with personal or work-related stress. An opinion survey of U.S. professional pilots suggested that the provision of employee assistance programs (a tertiary level countermeasure) would be the most effective approach to reducing the likelihood of drinking and flying (Ross and Ross, 1995). (Harris and Maxwell, 2001, p. 240)

Having read the above, it will be no surprise to the reader that a sample of professional pilots will be involved. It will also be obvious why this sample is to be used.

End of Chapter Summary

At the beginning of this chapter it was stated that the material covered in the previous five sub-sections should not be regarded as each representing a required paragraph or sub-section in the Introduction. They are simply issues that may need to be covered and justified to the critical reader. The most important thing is that the Introduction flows. It should set the scene and provide good reasons for all the decisions you made when designing and analysing your study.

When the reader gets to the aims and objectives of the work, described in the last paragraph of the Introduction (see Chapter 4) there should be no surprises at all. It should almost be self-evident from the preceding discussion why you did what you did.

Throughout writing this section it is important to keep track of your 'Big Message' and your bullet points. If what you are writing doesn't support the story of your research that you want to tell, *then cut it!* You need to learn to be ruthless when it comes to evaluating and editing your own work. You also need to keep re-visiting your Discussion section to check that you have introduced all the points that you make at the end of the paper. In practice, what usually happens is that there is a great deal of iterating between this section, the Discussion section and perhaps also the other parts of the Introduction (hence all those arrows in figure 1). I cannot emphasise enough that the manuscript must read as a 'whole', despite it being assembled in the slightly piecemeal manner that I am suggesting. All sections will continually require a little bit of revision throughout the manuscript construction process.

Writing the Introduction (Part Two): End of Chapter Checklist

Is there a clear description of the applied/practical background to the problem that follows on from the introductory paragraph (see Chapter 4)?	☐
Is there an analysis and synthesis of previous research describing the theoretical background to the study?	☐
Are all the references and issues talked about in the Discussion section introduced in an appropriate manner in the Introduction?	☐
Is there some justification (either explicit or implicit) of the methodology that will be used?	☐
Is there some justification (either explicit or implicit) of the dependent and independent variables that will be used?	☐
Is there some justification of the sample used in the study?	☐
Does the content of the Introduction compliment the 'Big Message' and the 'story' as described in the summary bullet points?	☐
Have you removed any material that is not required to deliver the message within the paper?	☐
Is the length and content of the Introduction in balance with the Discussion?	☐
Would you want to read what you have just written?	☐

So, you thought that you had finished? Think again. There are still quite a few bits and pieces to do, all of which are extremely important. These are all described in the next section so don't start relaxing just yet!

Chapter 7
Writing the Title, Abstract and Keywords

You may not think that these parts of the manuscript are of particularly great consequence, however once your paper is published they are potentially three of the most important aspects of your paper. They are also essential for getting your work published. In Chapter 4 it was pointed out that the reader of your final (published) paper will probably become aware of your work as a result of some sort of computer-based literature search process which involves:

- A keyword search undertaken on some abstracting and indexing service, followed by
- Reading the paper title (and maybe checking out the authors), followed by
- Reading the Abstract.

Only if the all of the above hurdles are cleared will your paper be downloaded (probably) and read. So, you need to get these right.

Title

Even after 25 years of authoring papers, I'm still rubbish at writing titles.[1] For some reason I find it exceptionally hard. As a result, most papers that I have had a hand in have incredibly dull and uninspiring beginnings. However, the upside of this is you always know what the paper is going to be about. And ultimately, this is not a bad thing. I once received a manuscript

---

1 Remember when writing any book or scientific paper, it is always important to instil confidence in your readers.

to review entitled *'genuflections, nictitations and anopsic ungulates'*. The content of the paper was concerned with the design of aircraft alerting systems, and how many systems (at the time) were difficult to understand and interpret: you couldn't work out what the warning was about. After about 30 minutes with various dictionaries (and a trip to the library was also needed to consult the serious dictionaries). I established that the title was basically a deliberately obscure way of saying 'a nod is as good as a wink to a blind horse'. A very clever title in the context of the manuscript, but if I was looking for a paper on aircraft alerting and warning systems I would probably have passed this one by.

If you want to be clever and try to produce an interesting title, Mark Young from Brunel University is probably one of the best exponents of the art. Along with a series of co-authors he has produced titles such as:

• Students pay attention! Combating the vigilance decrement during lectures.
• Crash dieting: The effects of eating and drinking on driving performance.
• Where do we go from here? An assessment of navigation performance using a compass versus a GPS unit.
• What's skill got to do with it? Vehicle automation and driver mental workload.

However, note that there is a pattern to his work. In all cases, the 'clever' bit is at the beginning of the title. The following part still informs the reader of the scientific content of the paper.

I'm afraid that my approach to producing a title is far more prosaic. The title usually either reflects what was measured, what was manipulated and the context of the study; or is a banal statement of what was produced. I also try and reflect the 'Big Message' in some way. This approach has produced such winners as:

• The effect of low blood alcohol levels on pilot performance in a series of simulated approach and landing trials.
• Pilots' knowledge of the relationship between alcohol consumption and levels of Blood Alcohol Concentration.

- Crosswind landing accidents in General Aviation: a modified method of reporting wind information to the pilot.
- Risk perception and risk-taking behavior of construction site dumper drivers.
- Development of a bespoke human factors taxonomy for gliding accident analysis and its revelations about highly inexperienced UK glider pilots.

It is worth examining carefully the formatting instructions for your target journal. Some publications have a restriction upon the number of characters in a paper's title. In this case characters always include spaces. This is often a 'hard' limit. With computerised submission of manuscripts it is impossible to exceed the prescribed number of characters.

When devising your title, *don't* include phrases such as 'a study into...'; 'an experiment...'; 'an investigation into...'. All of these things can be taken as read. Keep it short, simple, clear and to the point. The title should (a) attract the attention of your reader and (b) also give them a good idea about the content of the paper. Right from the start, they will begin to interpret everything that you subsequently say within the context set out by the title.

Abstract

You will be pleased to know that I am in much, much happier territory when it comes to writing the Abstract for a paper. After happening upon your paper as part of a literature search (via the keywords) this is your first major contact with the reader to give them an idea of the content of your paper and just how good it is. Indeed, even before your paper is published, the Abstract may be your first (unknowing) contact with your future referees. Many journals, when they approach a referee with an invitation to review a paper, will include the Abstract. This helps the potential reviewer to establish if the manuscript is within their area of expertise. Remember that first impressions count, so don't just throw together an abstract at the last minute. Spend time on it.

Right from the start you can 'sell' your paper, emphasising its strong points. Don't be too modest (but do try to be subtle):

This paper describes the development and initial validation of a **reliable and valid** multidimensional scale... . The results suggest that the scale shows both **content and construct validity**, being able to distinguish both between broad categories of vehicle and different models of vehicle within a particular category in a **consistent and meaningful manner**. (Harris, Chan-Pensley and McGarry, 2005 [emphasis added], p. 964)

Note the words that I have deliberately emboldened in the above extract. These were the key selling points about the final product of the research that I was trying to draw to the attention of the reader. The full Abstract can be found with the rest of this paper in Appendix 2.

Fortunately it is quite simple to produce a good Abstract. You will usually have about 150 words (maybe up to 200 – but certainly no more). So, you have one short sentence available for each major section of your paper (maybe two or three for the Discussion and Conclusions). Some Journals require sub-headings within the Abstract for each section. This can make your life a little easier and help to structure your work. The following Abstract, which requires this approach, is from a paper published in Aviation, Space and Environmental Medicine:

Introduction: Accidents caused by spinning from low turns continue to kill glider pilots despite the introduction of specific exercises aimed at increasing pilot awareness and recognition of this issue. *Method*: In cockpit video cameras were used to analyze flying accuracy and log the areas of visual interest of 36 qualified glider pilots performing final turns in a training glider. *Results*: Pilots were found to divide their attention between four areas of interest: the view directly ahead; the landing area (right); the airspeed indicator; and an area between the direct ahead view and the landing area. The mean fixation rate was 85 shifts per minute. Significant correlations were found between over-use of rudder and a lack of attention to the view ahead, as well as between the overall fixation rate and poorer coordination in the turn. *Discussion*: The results provide some evidence that a relationship exists between pilots' visual management and making turns in a potentially dangerous manner. Pilots who monitor the view ahead for reasonable periods during the final turn while not allowing their scan to become over-busy are those who are most likely to prevent a potential spin. (Jarvis and Harris, 2007, p. 597)

Before you start counting, it is 189 words There are several things to note. Don't try to include every reason and nuance from the paper in the Abstract. The idea is to give an overview, an impression, of what paper is about. Place greatest emphasis on the findings. Throughout Chapter 5 concerned with writing the Discussion section I kept saying that the greatest scientific value in any paper lies in the Discussion. This should be mirrored in the Abstract. The main interest of the reader is to establish what your conclusions were. The opening sentence of the Abstract sets out the context and says why the research is important. The sentence describing the methodology briefly describes how the major variables of interest were collected. However, two-thirds of the Abstract is devoted to the Results and Discussion. How were the data analysed; what was found and what implications did these results have, especially from the standpoint of the statement in the opening sentence?

The Aviation, Space and Environmental Abstract is quite long. Where you have fewer words available, consider cutting out the description of the Results section but do continue to maintain the greatest emphasis on the Discussion and Conclusions. The following 95 words are the Abstract from Harris and Maxwell (2001), in Appendix 1:

> A revision to the Joint Airworthiness Authorities' Operations Regulations now imposes a maximum Blood Alcohol Concentration limit of just 0.02% on U.K. pilots. Using a postal survey, opinions were elicited from 472 private and professional pilots concerning the effectiveness of various countermeasures to reduce the likelihood of drinking and flying. Punitive sanctions and tougher enforcement of the regulations were regarded as the most effective countermeasures, although offenders and professional pilots thought these actions less effective than private pilots and non-offenders. The results are discussed with respect to producing effective countermeasures specifically targeted at high-risk groups. (Harris and Maxwell, 2001, p. 237)

Note here that the Results section is hardly mentioned but the greatest proportion of the Abstract is still dedicated to the Discussion of the results and the findings. However, to find out more they will have to get the paper.

It can be quite a challenge to remove the final few words to get your Abstract down to the word limit. If so, get rid of all those 'airy fairy' terms – 'there is some suggestion that'; 'there was a trend towards'. Say 'it was'; and 'there was' (this is good writing, anyway). Concrete statements of fact always tend to be shorter. Words such as 'also' and 'additionally' can also(!) be removed. If there is any element of a list in the Abstract, there is no need to continually include the definite article. Once you have used a concrete noun of some kind (especially if it is composed of several words) consider replacing it with 'it' in the following sentence (as long as your meaning still remains clear). Again, as is the case with the title, when submitting your paper electronically you may run up against a 'hard' word limit in the submission software, so every word does count.

Finally, don't be too surprised if the content of your Abstract and the bullet points outlining the story in your paper have a great deal in common. They should have.

Keywords

With increasing use of computerised literature searches, keywords have become extremely important. These keywords may determine if your work is read and cited, or left on the electronic shelf to gather virtual dust. Ideally, these keywords would refer to sex, drugs, rock and roll, and fast cars (these terms should increase the hit rate for your paper[2]) but you also need to select five words (or short phrases) that best represent the main themes and findings from your work. They do, however, need to maximise the chances of your work being found in a literature search. The keywords may also be used by the Editor to select potential reviewers (regular reviewers' research interests are often held as keywords on a database held by the journal).

There is a very fine balancing act here, especially if you are working in a niche area. In such instances, the best keywords to describe the content of your paper may not enter the consciousness of even the most seasoned Human Factors researcher. They are simply search terms that would never normally be used. If you use a very common keyword, though (e.g. 'workload' or 'accident') then your paper may end up being just one of many in the final search results.

The best option is to use the most appropriate combination of the common terminology to describe your work. For example, use 'situation awareness' instead of the less common 'situational awareness' or 'workload' instead of 'taskload' (even if your study is more concerned with the latter). Don't worry about differences in UK and US spellings. Most search engines are smart enough to accommodate these differences. Many search engines also utilise

---

2   Maybe I am thinking like a man, here.

a thesaurus of some kind. If you are only allowed five or six keywords and you are working in a particularly esoteric area, only allow yourself a couple of words relating directly to the niche area to describe the content of your paper. Make sure the remainder of your keywords are in more common Human Factors usage. For the Bohm and Harris (2010) paper which was concerned with the 'niche' area of dumper truck driving, of the five keywords allowed, three addressed reasonably common Human Factors Concepts ('risk perception'; 'hazard awareness' and 'accidents') while the other two used the more abstruse (but more applicable) terms 'construction site dumper' and 'construction site safety'.

If your study builds upon the work of previous authors, what keywords did they use? What keywords did you use to find their work in your literature search? Keywords count. As we will see in a short while, there are prizes for having your paper downloaded.

Writing the Title, Abstract and Keywords: End of Chapter Checklist

Does the Title clearly reflect the content of the paper?	☐
Does the Abstract set out the context of the work in its first sentence?	☐
Does the Abstract use only one sentence to describe the Methodology employed?	☐
Is the largest proportion of the Abstract dedicated to the findings from your work?	☐
Are the 'Big Message' behind your paper and the summary bullet points describing its story recognisable in the Title and content of the Abstract?	☐
Have you selected five, appropriate keywords?	☐
Would you be able to find your paper if you searched for it?	☐
Would you want to read what you have just written (the Abstract)?	☐

That just about does it for the actual writing of the manuscript, unfortunately there is still some hard work to do before the paper can be submitted to your journal of choice. It is also the next things that new authors most often get wrong.

Interlude III:
Some Different Types of Paper

This book has concentrated solely on what may be labelled, full 'scientific' papers. However, there are various other categories of paper that are worth considering. With the exception of the final couple of types of paper mentioned in this section, all the others described can be constructed in a manner similar to that expounded in this book.

Practitioner Papers

These papers describe the application of scientific principles, rather than the generation of 'new science' *per se*. They are often slightly shorter than scientific papers and may cover a wide range of topics such as structured comment and analysis of incidents and accidents; papers covering innovative applications of scientific principles and reviews of best practice in industry. These manuscripts are often written in a more accessible (somewhat less technical) style and may be more liberally illustrated with diagrams and pictures. Depending upon journal policy, these papers may not be quite so rigorously reviewed as the full scientific articles.

Research Notes/Short Communications

Some journals invite shorter papers, typically in the region of around 3,000 words (including references) and containing no more than one or two figures, a couple of tables and a limited number of references (maybe 10–12). These are intended to be faster communications that can bring developments to the notice of readers quicker than full papers. However, with the advent of the web-based pre-publication of papers (e.g. the Taylor and

Francis *i*First early, online publication system – see http://www. informaworld.com/smpp/ifirst) there is becoming less need for this type of article compared to the time when all journals were published only in a paper format). Many journals still invite this type of submission, though (e.g. *Aviation, Space and Environmental Medicine*).

Commentaries

A few journals invite commentary papers. Commentaries are brief essays that set out a personal opinion or perspective on a contemporary topic. Such manuscripts are usually quite short (about 1,000 words) with no tables or figures and only a few references.

Review Papers

If you want to increase your personal citation rate the best way to do this is to write a review paper. It has been assumed throughout this book that the manuscripts being prepared are based upon extant research (for example in the form of a technical report or a thesis). However, before doing any major piece of research a literature review is essential. In some cases (e.g. in the case of a PhD thesis) these can be a major piece of work in themselves. Critical review articles or meta-analyses covering a particular topic of interest are welcomed by many journals (they help increase their Impact Factor). Some journals (such as *Theoretical Issues in Ergonomics Science*) are dedicated almost entirely to this type of paper.

Nevertheless, any review paper must have a point to it. It should not just be a historical overview of a particular area. There should be analysis and synthesis of the topic and a critical review of the material but from a specific applied or theoretical perspective. It should not be an extended Introduction (which is a lead in to the study that follows). The material in a review paper should be an end in itself and it should develop the theoretical underpinning of a research area. If you want to know 'how to do it' I would suggest that you investigate many of Neville Stanton's contributions to *Theoretical Issues in Ergonomics Science.*

Chapter 8
Formatting and Submitting your Manuscript

This again, is one area where many new authors reveal themselves. Many submitted manuscripts do not adopt the format required by the journal. Note that manuscripts do not look anything like the corresponding final paper. The required format for manuscripts facilitates anonymising the contribution and ultimately assists in the final editing and production of the paper as it appears in the journal. To be honest, though, the format of most manuscripts does nothing to promote the reading and review procedure.

As an Editor, when a new manuscript arrives and it looks like a thesis or technical report (or even in some cases, I have seen instances where the contributor has taken the trouble to make it look exactly like how the paper would appear in the journal) you immediately think 'new author'. And then you start looking at the manuscript quite hard to evaluate its content. As you will find out in the following chapter, submissions that do not conform to the required format will not get far through the reviewing process. However, the effect has also immediately been to create a negative impression. If you prepare the manuscript exactly as it is required by the journal, the first impression is that an experienced author has had a hand in its creation. First impressions count.

Formatting Your Manuscript

Although formats differ slightly from publication to publication, there is a great deal of common ground. All manuscripts tend to be double spaced throughout; usually have tables appended at the end of them (*not* where they would appear within the main text) and figures are appended as separate files. The position where a table or figure should appear in the text is simply marked with a comment, such as:

INSERT TABLE/FIGURE X ABOUT HERE

The main differences between journals in their manuscript requirements are usually in the format for headings and for references (both callouts in the text and the format of the full references at the end of the paper). However, before getting into these relative subtleties, don't forget the most obvious formatting requirements: what size paper is required? A4 or US letter (10 × 8 inches). Is there a word limit? If so, what is it and is the manuscript within this bound? There is usually *a little* leeway here but only 10%, or so (maybe…). Get these simple things right first.

In general, most journals require something like the following:

- Margins: 3 cm (or 1 inch) on all sides (top, bottom, left, right).
- Font: 12-pt. if a proportional font, usually Times New Roman font; if a non-proportional font is required, often Courier is specified.
- Line spacing: Double-space throughout the manuscript.
- Alignment: Flush left (*not* right justified). First line of every paragraph should be indented (except the one following a heading).
- Pagination: The page number should appear somewhere on every page.
- Running head: This is a short title appearing at the top of each page, as a rule not exceeding about 50 characters.

It is not the object of this book to describe all the common formats for manuscripts. Simply go and check the journal's web site. Many journals also have abridged formatting instructions on the inside back cover. APA formatting instructions are quite typical and widely used (but not exclusively) in Human Factors. However, I must caution you (slightly) about the Sixth Edition of the APA Publication Manual. It is a fantastically comprehensive book on the required format for APA manuscripts and the suggested style for their preparation: the absolutely authoritative work on this topic. However, it is over 260 pages long and is not a 'light read'. Fortunately, the APA also produces a free web-based video tutorial about its required style, available at www.apastyle.org/learn/tutorials/basics-tutorial.aspx. This is a great introduction.

Most formats require that the title page of the manuscript, with the authors' name(s), affiliation and contact details comes first or is kept as a separate file (when using web-based submission systems). This is to facilitate the anonymisation of the paper when it is sent to the reviewers (see Chapter 9 describing the manuscript review process). The following page is usually confined to the Abstract (and maybe keywords). No authors' names should appear on it. Again, this is to facilitate the reviewing process and ensure that the contributors' name(s) remain hidden from the reviewers. Using a web-based submission system the Abstract will probably be lodged as a separate step in the process (and hence a separate file). Only on the third page will the manuscript itself start. Again, this will commence with just the paper title: no authors' names. This is the part of the submission that will be sent to the reviewers (along with any tables and figures attached). In general, it is your responsibility to remove all potentially identifying information from your submission by replacing your name (or institution name) with some kind of neutral placeholder. If the manuscript is accepted for publication, these items will (of course) be reinstated.

Typically, any formatting system will have at least three (often more) levels of heading. The top three levels of heading when using APA style are:

- Level 1: Centred, boldface, using uppercase and lowercase letters (capitalise each word). First paragraph of the section starts on the following line.
- Level 2: Flush left, boldface, using uppercase and lowercase letters (capitalise each word). First paragraph of the sub-section starts on the following line.
- Level 3: Indented, boldface, sentence case. Heading ends with a full stop. Sub-section begins on the same line immediately after the heading.

Level 1 headings will be used for all the major sections (Introduction, Method, etc.). Try not to use too many levels of heading. Although the judicious use of headings can help to provide the reader with an appreciation of the manuscript's structure and present useful signposting of the issues, too many headings at too many levels can also break up the paper unnecessarily and make it disjointed. I would suggest never going beyond three levels of heading in a scientific manuscript.

All source material *must* be acknowledged in the manuscript. There are various common ways of doing this, but the two most common are:

- By citing author(s) and date(s) in the text. When a citation has one or two authors, all authors are included every time that the material is referred to. When a source has three or more authors, the full list should be used the first time the source is cited. Subsequent citations should simply use the first author's surname followed by '*et al.*' and the date. Common examples of this format of callout are the APA style and the Harvard referencing system.
- By including a number which refers to a full reference included in the References section. In this case the authors' names are not necessarily included in the text. The convention for numbering may either be by the order in which the reference first appears in the manuscript (e.g. the Vancouver system, which is commonly used in medical and engineering journals) or by the alphabetical order of the authors (as used in some engineering journals, for example *Aerospace Science and Technology*).

The convention for citing references in the text may require a small degree of re-phrasing of certain aspects of your manuscript. For example, the following extract is from Payne and Harris (2000):

> Gibson (1995) expanded on this definition to include the ease with which the pilot could compensate for the disturbing effects of the environment. (p. 344)

This extract uses the APA style for citing a reference. However, when using an alternative style, such as the Vancouver system which numerates references in the manuscript, this sentence may be re-written, thus:

> This definition was expanded to include the ease with which the pilot could compensate for the disturbing effects of the environment (1).

Quotations should always be made obvious, by italicising or enclosing in quotation marks, and followed by a callout to the author, year, and page number. Longer quotations (over about 50 words) should be italicised, indented and in block format.

Another major difference between these referencing systems occurs in the manner in which the full article is presented in the References section. It is beyond the scope of this book to provide a full guide to all the different referencing conventions but here is a *very* brief guide to the three most common formats (APA; Harvard and Vancouver). If you are in any doubt, go and find a copy of your target journal and follow *exactly* the formats used to present the references in it.

APA Style

Journal Article

Ebbatson, M., Harris, D., Huddlestone, J., & Sears, R. (2008). Combining control input with flight path data to evaluate pilot performance in transport aircraft. *Aviation Space and Environmental Medicine, 79*, 1061–1064. DOI: 10.3357/ASEM.2304.2008

Book

Martinussen, M. & Hunter, D. (2010). *Aviation Psychology and Human Factors*. Boca Raton FL: CRC Press.

Book Chapter in an Edited Book

Jorna, P. G. A. M., & Hoogeboom, P. J. (2004). Evaluating the flight deck. In D. Harris (Ed.), *Human factors for civil flight deck design* (pp. 235–274). Aldershot, UK: Ashgate.

Note that Digital Object Identifiers (DOIs) should now be included at the end of the reference (if available). Be particularly careful with the use of ampersands, parentheses, italicisation and spaces.

Harvard Style

Journal Article

Ebbatson, M. Harris, D. Huddlestone, J. and Sears, R., 2008. Combining control input with flight path data to evaluate pilot performance in transport aircraft. *Aviation Space and Environmental Medicine, 79,* pp. 1061–1064.

Book

Martinussen, M. and Hunter, D., 2010. *Aviation Psychology and Human Factors.* Boca Raton FL: CRC Press.

Book Chapter in an Edited Book

Jorna, P.G.A.M. and Hoogeboom, P.J. (2004). Evaluating the flight deck. In: D. Harris, ed. *Human factors for civil flight deck design.* Aldershot, UK: Ashgate, pp. 235–274.

Vancouver Style

Journal Article

Ebbatson M, Harris D, Huddlestone J, Sears R. Combining control input with flight path data to evaluate pilot performance in transport aircraft. Aviat Space Environ Med 2008; *79:* 1061–4.

Book

Martinussen M. Hunter D. *Aviation Psychology and Human Factors.* Boca Raton FL: CRC Press, 2010.

Book Chapter in an Edited Book

Jorna PGAM, Hoogeboom PJ. Evaluating the flight deck. In: Harris D, ed. *Human factors for civil flight deck design.* Aldershot, UK: Ashgate; 2004: 235–74.

Note that when using the Vancouver style, journal names are commonly truncated by using the NLM (National Library of Medicine) *Index Medicus* abbreviation for their title. The full list of journal title abbreviations can be downloaded from ftp://nlmpubs.nlm.nih.gov/online/journals/ljiweb.pdf.

It is a good idea to hold your references using some bibliographic management software (such as Endnote™ from Thomson Reuters or RefWorks™ from ProQuest) which allows you to create a personal database and automatically generate references in a variety of standard formats. This will save you a great deal of time when producing the final list of references and will help to ensure that they are presented correctly. Furthermore, the more references that you hold and the more that you write, the more time that it will save you in the future. It will also assist you when re-numbering any references in the text if you are using the Vancouver referencing style (for example, if you have to delete, insert or move a citation). Doing this manually is a nightmare! If you are doing this manually, wait until you have completed the manuscript once and for all, and you are happy with it before you start numbering citations in the text. And then do it with great care (I usually do it with someone, who checks references off as I call them out to make sure that nothing is put in twice or omitted).

Take the time to format your manuscript to the exact requirements of the journal. You don't want to look like a novice.

Submitting Your Manuscript

The majority of journals now use a web-based system for submitting manuscripts; logging their progress in the review process, and for returning reviewer comments to the authors. There are two common software systems: Editorial Manager™ from Aries Systems (see http://www.editorialmanager.com/homepage/home.htm) and ScholarOne Manuscripts™ (see http://scholarone.com/products/manuscript/). Some journals still use a simple e-mail exchange with the Editor. Nevertheless, there is one critical document that you still need to write: the submission covering letter.

Covering Letter

This is not just a nice letter to the Editor: the submission covering letter should attest that:

- The manuscript has not been published previously (or that it is simultaneously under consideration for another journal).
- The manuscript is being submitted with the agreement of all authors.
- There are no known conflicts of interest; and
- All participants and their data were treated in accordance with the appropriate ethical guidelines.

For example, this is a basic template for a covering letter that I use for most of my manuscript submissions:

Dear [INSERT EDITOR'S NAME]

Please find attached a manuscript entitled '[INSERT MANUSCRIPT TITLE]' by [INSERT LIST OF AUTHORS]. We would be grateful if you would consider this paper for publication in [INSERT FULL JOURNAL TITLE].

This is an original manuscript not being submitted elsewhere for publication. We believe that there are no conflicts of interest (financial or otherwise) that impinge on the quality or the impartiality of the data obtained. All authors have agreed the content of this version of the manuscript.

The treatment of the people and the data in this study conformed to the ethical standards required by the [INSERT COUNTRY] professional body, [INSERT NAME AS APPROPRIATE HERE].

We look forward to receiving the comments from the referees.

Yours Sincerely

[YOUR NAME]
Enclosures: [LIST IN HERE]

Electronic Submission

Having done this you can then either e-mail your submission to the Editor, or more likely, log onto the journal manuscript management system and upload it. This is usually done via the journal's home page. If this is your first time you will need to create an account (naturally) which will involve creating yet another username and password. You will need to enter various basic personal details, but you should only need to do this for the first time that you submit a manuscript to the journal.

When it comes to submitting the manuscript itself, you will probably be required to enter the title of the paper; authors (in agreed order, including their affiliation and contact details); keywords and Abstract into separate electronic fields. This helps in the process of anonymising the manuscript for review. The word processor file containing the manuscript will then need to be uploaded, followed by any files containing figures. These should be in the required file format (frequently the software will refuse to recognise anything other than acceptable formats) and file names should refer explicitly to the figure number and authors. At some point the submission letter will also need to be uploaded.

After doing all this, the manuscript will then automatically be converted into a .pdf file and returned to you for your approval. Look at it very carefully. If you use a facility in your word processing software that automatically creates indexes from headings, or if you use software for storing and inserting your references, check that these links have been recognised and inserted at the appropriate points in the final manuscript. If I had a pound for every time I have received a manuscript with '*ERROR! Bookmark not defined*' somewhere in it, I would still be a very poor man but at least I would be able to buy a new (reasonably priced) armchair in which I could while away my old age. The final .pdf created file will be the one circulated to your referees.

In general, I find that when preparing a manuscript I have greatest success when using the fewest possible number of automatic features on my word processing program. Don't set up automatic level 1, level 2 headings, etc. Simply embolden and position the text yourself (as per the journal formatting requirements). I have found that this approach is the most reliable way of creating error-free .pdf files for manuscript submission.

The submission software will also make some cursory checks on the quality of your figures. It will not be able to check automatically the font sizes used, appropriate shadings, etc. but it will assess their resolution. For refereeing purposes it is usually acceptable to submit lower resolution figures than would ultimately be desirable (you normally need a resolution of at least 600 dots per inch) but do not be surprised if you get a message back telling you that your figures are of low quality and will require re-drawing if the manuscript submission successfully completes the review process.

So, check the .pdf file containing your final manuscript; check any error/advisory messages concerning the quality of your figures; check that you are happy with your title, abstract and keywords; check the order of authors (and who is designated as the corresponding author) and check the .pdf file with the submission letter. If everything is OK you can approve the manuscript it and hit the 'submit' button. You have finally done it!

Now it will just need revising after receiving the referees' comments...

Formatting and Submitting Your Manuscript: End of Chapter Checklist

Is the manuscript formatted to exactly the requirements specified by the journal?	☐
Is there a separate title page with authors' contact details on it?	☐
Is there a separate page containing the Abstract?	☐
Are figures saved as separate, appropriately named files and in the correct format?	☐
Are all references present (which are called out in the text) and are they presented in the correct format?	☐
Have you created a user account for electronic submission (if applicable)?	☐

Have you written a covering letter for your submission?	☐
After you have submitted your initial word processor file, have you checked the pdf file created for accuracy and omissions?	☐
Are there any immediate problems with the quality of the figures that require action before the manuscript can be submitted?	☐
Would you want to receive this manuscript to review?	☐

Interlude IV:
Top 10 Mistakes

Every popular book nowadays must have a 'top 10' list in it somewhere (or should that be a bottom 10 in this case)? So, in no particular order, here is my list of the most common mistakes made when preparing a manuscript for submission to a Human Factors journal:

1. Producing a paper that does not have a coherent message all the way through it, which is often a product of trying to say too much.
2. Too much detail in the Introduction: lots of irrelevant references (the *'I have read it, so I'm going to include it'* phenomenon, or *'the referees will think that it is a good paper, I have used a lot of references'* fallacy)!
3. Lots of simple description of previous authors' works in the Introduction and not enough critical analysis and synthesis within the context of the research being described in the paper.
4. Poor description of the Method: an intelligent reader could not reproduce the methodology from the detail provided (failing the 'Delia Smith' test).
5. Poor production of figures in the Results section (font too small; lots of use of colour in figures in a journal that will be printed in black and white; poor use of shading; figures too complex to reproduce on a limited page size; use of PowerPoint™ and Excel™ graphs and figures that won't reproduce or re-size properly – should use .TIF; .GIF; .BMP or .AI, etc.) and many figures irrelevant.
6. Poor production of tables in the Results sections (many using statistical package variable names; too complex; font too small; format doesn't reflect the hypotheses tested, etc.).

7. Lots of direct use of output from statistical packages, rather than presenting the results of the statistical analyses using the accepted abbreviated form within the main text.
8. Not enough detail included in the Discussion: no explanation and does not link the Results to the material in the Introduction (and this is the important bit – science is all about the explanation and interpretation of phenomena).
9. Failure to follow the required format for the presentation of a manuscript to a journal.
10. Failure to consider everything that they have written from the point of view of the reader, particularly assuming too much knowledge on their part and presupposing that they have a perfect memory for what has gone before.

Chapter 9
The Manuscript Review Process

One you have submitted your manuscript to the journal, it will be subject to the peer review process. The objectives of this chapter are to give a brief outline of this process, the role of the Editor and to describe some of the things that your referees will be looking for in your paper. How to answer their criticisms of your paper will be covered in the next chapter. However, there are two vital points to note here. Firstly, the purpose of the peer review process is to improve the quality of papers published in the journal (it is not there to frustrate authors). Secondly, it is one thing to write a good scientific paper of the required quality for publication; it is quite another thing to finally get it published.

Peer Review Process

The peer review process proceeds as follows:

- The manuscript is submitted to the Editor by the paper's nominated corresponding author.
- The Editor then initially scrutinises the paper, and if it is potentially suitable for publication in the Journal, processes it, and sends it out for review to two, three (or sometimes more) suitable referees.
- The referees then critically read the manuscript, comment on it, and return their reviews to the Editor, along with their initial recommendation for publication.
- The Editor reviews the referees' comments and recommendations, collates them, and passes them back to the corresponding author along with his/her decision.

This process can be repeated several times for a manuscript (depending upon journal policy) before it is either accepted for publication, withdrawn by the authors or finally rejected by the Editor.

There are three main approaches to peer review. The norm for most Human Factors journals is to employ the 'double blind' peer review process. This is held up to be the highest integrity peer review process. When using this approach, the authors submitting the manuscript are not aware of the identity of the reviewers of their paper. Similarly, the identity of the author(s) is concealed from the referees in case this may bias their reviews. To preserve the integrity of this process, in all correspondence (or 'phone calls) the authors will only usually deal with the Editor of the journal. Some Editors may delegate the review of manuscripts in a particular scientific area to an Associate Editor with greater expertise in that domain. However, in such an instance, the authors may not even know the identity of the Associate Editor responsible.

There are some criticisms of the 'double blind' peer review process. Seasoned reviewers and academics can often recognise other prolific authors from the context of the work, their scientific approach, writing style, or most easily, from the number of self-citations in the References section. In such a case it is not a truly blind process. It has also been argued that using this review method an unscrupulous referee can attempt to thwart the publication of findings that contradict their own particular position on an issue with impunity (a good Editor should stop this). A further criticism of this approach concerns the delay it introduces in the publication process. Some journals attempt to process manuscripts within 30–50 days, however it is more typical for the double blind review process to take considerably longer (often over 100 days).

One alternative is the single blind peer review, in which case the author's identity is known to the reviewers but the reviewers' identity is hidden from the author(s). This is used by some journals (although it is much less common than double blind peer review). One justification for this approach is that totally double blind peer review is impractical (for the reasons outlined previously). There should be no biases in the single blind process as a reviewer should declare any conflicts of interest to the Editor when the names of the manuscript author(s) become known to them. Any conflicts should prohibit the potential referee from reviewing the paper. However, the Human Factors scientific community is quite small; many colleagues are friends, and speaking as a journal Editor, good reviewers are very hard to get. Good reviewers also tend to be good authors.

The single blind peer review process is also seen as a potential way of promoting open peer review. In open peer review the identities of both the author(s) and those reviewing the manuscript are known to each other. While appearing to avoid any criticisms of bias by promoting transparency it has also been suggested that in open review, more junior researchers are reluctant to criticise the work of their senior, more esteemed peers. In some cases where it has been trialled there was a marked reluctance among referees to offer open comments. Many journals that operate a single blind peer review process give the reviewer the option to make their name known to the author on completion of the review, hence effectively making it an open review process.

The Editor

After your manuscript has been submitted to the Journal of your choice, its first port of call will be the Editor's desk. The Editor is responsible for producing a high quality academic periodical. To do this they need to attract good papers from good authors, subject them to a high standard of constructive critical review and present them to their readership in the best way that they can. They select the best manuscripts that they have available. For the Editor, quality is everything.

If you are unlucky, or if you have not done your homework correctly, you can have your manuscript returned to you (or even rejected) by the Editor at right at the beginning of the review process without it even being sent to the referees. From my experiences as an Editor, about 25% of all manuscripts submitted fail at this stage. As I have said, good referees are very hard to find, so I treat them as a precious resource that I try not to overburden. Removing the detritus early is one way of doing this.

The most common reasons for getting a 'desk rejection' from the Editor are:

- Manuscript not prepared to the required format/standard. The scientific content might be satisfactory but its presentation is poor or inappropriate. In such a case the manuscript can easily be re-worked and re-submitted, however, you have already incurred the wrath of the Editor. You have demonstrated that you are not particularly familiar with his/her journal and that you cannot follow the simple instructions for formatting a manuscript (or perhaps you choose to think that these rules do not apply to you)!
- Manuscript not within the remit of the Journal. The science in it may be satisfactory but the subject matter is outside the scope of papers normally dealt with by the journal. If you are unsure whether your material will be acceptable to the journal, contact the Editor first. Submitting material outside the remit of the journal again demonstrates that you are not particularly familiar with his/her journal. In this case there is no point in re-submitting your work after it is rejected.
- The manuscript is simply of very poor scientific quality (again, no point in re-submission).

If you are successful at this point, the Editor will then anonymise your manuscript before sending it out to be refereed. One of the steps in doing this is to remove the first page of the manuscript (either physically if it is 'real' paper or electronically from the submitted file(s)). This is why all the authors' personal details are required to be on a separate page in the manuscript formatting instructions. The same applies to acknowledgements. The manuscript will then be assigned a unique number, which should be used in all correspondence.

From looking at the content of the manuscript and the methodological approach used, the Editor (or Associate Editor) will then approach two or three potential referees to review the manuscript. These may be selected on the basis of the keywords that you have provided to describe your work. To do this, the Editor will usually e-mail the Abstract of the paper to them to establish if the work is within their area of expertise (as well as establishing if they have the time and are willing to complete the review). This is why manuscript formatting instructions usually require the Abstract to be on a separate page. Every time you don't follow the required manuscript format you involve the Editor in more work. Remember, first contact with your referees will be via your Abstract. This is precisely why you should put a bit of effort into writing it and why you need to present your findings in a positive light in the last sentence.

The reviewers will most likely be members of the Editorial Board, however this is not exclusively the case. If they agree to review the paper the Editor will then send them the anonymised copies of your manuscript and usually ask them to return their comments and recommendations within 28 days.

On receipt of the reviewers' comments the Editor will read and evaluate them. As you will soon see, the Editor does not expect to see the same things appearing in all the reviews (in fact this is quite rare), however s/he will expect to see some commonality of opinion concerning the overall quality and applicability of the manuscript. Where there is a big discrepancy in the reviews received, especially in the final decision, the Editor will usually send the manuscript out for a further opinion. This is where many delays occur.

The Editor is responsible for making a final decision concerning the suitability of the manuscript, based upon the recommendation of the reviewers. The usual decisions are either:

- Accept
- Accept (minor corrections)
- Accept (major corrections and re-review)
- Reject

The meaning of 'Accept' is obvious (but is a very rare outcome)! This means the manuscript is suitable for publication with no revisions whatsoever. 'Accept (minor corrections)' usually refers to a manuscript which needs a number of small revisions but these are confined to stylistic and typographical corrections (and maybe some very small clarifications). These should be returned to the Editor, who is responsible for verifying them and approving the final paper. In this instance there is no requirement for the manuscript to be returned to the reviewers. 'Accept (major corrections and re-review)' usually means that the content of the manuscript is probably acceptable for publication in the journal, but there may be major omissions; mistakes; a lack of clarity or some re-analysis or re-interpretation of the results is required. The manuscript needs re-review. In this instance the manuscript can be returned to the authors several times if the initial revisions are not up to the standard required. Some journals have a policy on the number of revisions that will be allowed before a manuscript is either finally accepted or rejected. Quite often this is one round of Major Revisions followed by one round of Minor Revisions. All of these things are at the ultimate discretion of the Editor, though. The 'Reject' decision may sometimes be split into two categories: 'Reject' with no opportunity to re-submit and 'Reject' with the potential opportunity to submit the paper in the future as a completely new manuscript (after it has been completely re-worked). I only tend to use the former in the case of a manuscript which is based upon research that has a fundamental flaw in its design, meaning that any data in it are of dubious validity.

This decision will then be communicated to the corresponding author of the paper, along with copies of the reviews from the referees. Discussing the final decision with the Editor (in the case of a negative outcome) is usually a pointless exercise. If you don't like the decision, submit the paper to another journal! As an Editor I have been on the receiving end of some unpleasant e-mails and 'phone calls from disappointed authors who have had their work heavily criticised or rejected. I can assure you that in these cases, such communications did little to enhance the chances of these manuscripts being published.

The Reviewers

The Editor will send your manuscript to at least two reviewers for critical comment. These will be independent experts in the area. The reason for using two (or more) reviewers is not necessarily to look for concordance in their comments and criticisms: in fact, quite the reverse. Different people see different things in a manuscript (both weaknesses and strengths). Using more than one reviewer allows a greater number of perspectives to be obtained and fed back to the authors and Editor. So don't be surprised when your two reviewers produce very different reports on your paper. This is not a bad thing.

What are the reviewers looking for in a manuscript? Catherine Daily and Albert Cannella produced the following guidelines for referees when evaluating manuscripts submitted to the Academy of Marketing. However, they apply equally as well to manuscripts from any discipline.

- Introduction
    - Is there a clear research question, with a solid rationale behind it?
    - Is the research question interesting?
    - After reading the introduction, did you find yourself motivated to read further?
- Theory
    - Does the submission contain a well-developed and articulated theoretical framework?
    - Are the core concepts of the submission clearly defined?
    - Is the logic underlying the hypotheses persuasive?
    - Is extant literature appropriately reflected in the submission or are critical references missing?
    - Do the hypotheses (or aims and objectives) flow logically from the literature presented?
- Method
    - Are the sample and variables appropriate for the hypotheses?
    - Is the data collection method consistent with the analytical technique(s) applied?
    - Does the study have internal and external validity?
    - Are the analytical techniques appropriate for the theory and research questions and were they applied appropriately?
- Results
    - Are the results reported in an understandable way?
    - Are there alternative explanations for the results, and if so, are these adequately controlled for in the analyses?

- Contribution
    - Does the submission make a value-added contribution to extant research?
    - Does the submission stimulate thought or debate?
    - Do the authors discuss the implications of the work for the scientific and practice community?

It is worth looking at these bullet points and doing a self-review of your manuscript before submitting it. Of particular note is the last major bullet point 'Contribution'. It is not enough just simply to do a good, well-founded piece of research. Somehow, in some way, it has to be new. You need to spell out your novel contribution to the reviewers. Tell them what it is – make their minds up for them!

Most Human Factors journals do not use extensive structured check lists for reviews despite their common usage in other disciplines. Use of a more structured approach can aid both reviewers and authors, ensuring a comprehensive evaluation of all major points within the manuscript (and hopefully obtaining a greater degree of agreement from the referees on key aspects). The use of checklists should be balanced, however, with the opportunity for referees to provide extensive, qualitative feedback.

In 2010/11 I was involved (as part of EAAP – European Association for Aviation Psychology) in the development of a new Human Factors journal: *Aviation Psychology and Applied Human Factors*. Right from the start a semi-structured review *proforma* was developed for manuscript reviewers. An early version of this review checklist is reproduced (in full) in Figure 9.1. One objective when producing this instrument was to ensure that not only was the content of each section of the manuscript explicitly considered (and commented on) by each reviewer, they also had to take into account the overall structure, flow and coherence of the paper. You will notice that many of the points in this review form have also been raised in the end of chapter checklists. This is not a coincidence! When reviewing, referees are also free to append other documents to their reviews (and many still choose to use an unstructured review). In some cases reviewers now use the facility in Word™ to append notes to the original manuscript with their comments.

Manuscript Number:

Manuscript Title:

Does the title describe the content of the paper? | *Select*

Comments on the Title:

Does the content in the introduction adequately describe the theoretical and practical background of the paper? | *Select*

General Comments on the Introduction:

Are there any major omissions in the introduction? | *Select*
Is the introduction well structured? | *Select*
Does the introduction end with a clear statement of the aims and/or objectives of the study? | *Select*

Further Comments:

Does the content of the material in the methodology section adequately describe the data collection techniques applied? | *Select*

General Comments on the Methodology Section:

Are the sample characteristics adequately described? | *Select*
Is the methodology appropriate? | *Select*
Are the data collected appropriate for the study? | *Select*
Are all dependent and independent variables adequately described? | *Select*
Is the procedure adequately described and easy to follow? Could you replicate the procedure from the description supplied? | *Select*
Is the method section well structured? | *Select*

Further Comments:

Does the content in the results section support the stated aims and objectives of the paper? | *Select*

Figure 9.1   Review instrument used by referees contributing to *Aviation Psychology and Applied Human Factors*. Note that the boxes for free text expand as text is entered. The reviewer in not limited in the number of comments that they can enter in each text box.

General Comments on the Results Section:

Is any treatment of the data clearly described? | *Select*
Are all the necessary data presented? | *Select*
Are the data presented clearly and appropriately (but without repetition)? | *Select*
Are the data analysed appropriately? | *Select*
Are results interpreted correctly? | *Select*
Is the results section well structured? | *Select*

Further Comments:

Does the content of the discussion section support the stated aims and objectives of the paper? | *Select*

General Comments on the Discussion Section:

Are the results discussed within the practical and theoretical context described in the introduction? | *Select*
Is there a clear and explicit link to the material in the introduction? | *Select*
Is there an appreciation of any potential methodological limitations to the study? | *Select*
Is the Discussion section well structured? | *Select*

Further Comments:

Does the paper end with clear conclusions and recommendations that relate directly to the aims and objectives in the introduction? | *Select*

General Comments on Conclusions and Recommendations:

Now you have read the paper, does the abstract adequately describe and summarise the content of the Manuscript? | *Select*

Figure 9.1 continued

General Comments on the abstract:

General Comments on the paper:

Is the content of the paper within the remit of the Journal?		*Select*
Which section of the Journal should the paper be published in?		*Select*
Are the references presented in the appropriate (APA) format?		*Select*
Is the standard of English appropriate for an International Journal?		*Select*
Was the paper easy and enjoyable to read?		*Select*

Final General Comments on the Paper:

Decision

In your opinion, should the manuscript be: *Select*

Thank-you

Figure 9.1 concluded

Collecting comments in this structured manner also allows for easier archiving of reviewer feedback. Most journals also rate the reviewers. In the same way that there are good and bad authors, there are also good and bad referees. Good Editors pay careful attention to the quality of feedback provided to authors by the referees that they appoint.

A Closing Note

Always remember that the Editor, Associate Editors and the reviewers are all volunteers who give their time up to run the journal for very little reward (maybe just a free subscription and some expenses). Anything that makes their life a little bit easier when appraising your manuscript is appreciated. Treat them with respect and be grateful for their feedback – you are getting their expert opinions for free. A good reviewer working with a good Editor is always looking to enhance the quality of your paper. They are not trying to give you a hard time. And bear in mind that it is a lot less effort (from both the Publisher's and the Editorial Board's point of view) to say 'no', than to say 'maybe' or 'yes'.

Chapter 10
Responding to Referees' Comments

If you have received the comments on your manuscript and it has not received an out and out rejection from the Editor and referees, then to get it published you will probably have to respond to the latter's criticisms of your paper. At this point try to bear in mind that one of the major roles of the referees is to improve your paper. They are not pursing a personal vendetta against you or trying to sink your paper without trace. I mention this now as this is a common reaction on opening the reviewers' comments for the first time.

Even now, I still sometimes take the referees' comments personally, even when they are very nicely written. You can't help it – someone has attacked your baby! A common first reaction tends to be that they know nothing and don't really understand what I am trying to say in the paper (which is an interesting comment in itself...). So, for many years I have adopted the following strategy. Receive the reviewers' comments: open them: read quickly: fume quietly to myself for a while: put the comments away for at least two days (*do not touch*): re-read comments and now decide that they are actually not that bad – the majority of the criticisms are very fair and are quite easy to address: fix the paper...

A re-occurring theme throughout this volume has been that it is one thing to produce a well written manuscript but it is quite another thing to get it published. One tactic to help get a paper published is to make the Editor's and referees' lives as easy as possible. This is exactly what I try to do when responding to comments and implementing the reviewers' suggestions. It does not matter what the overall outcome of the review is ('accept with minor corrections'; 'accept but with major corrections and re-review' or even 'reject, but with opportunity to resubmit the manuscript') the approach I advocate for undertaking the revisions is exactly the same. It is merely a question of scale.

However, the first step is simply to e-mail the Editor to acknowledge the receipt of the comments, thank the reviewers and advise him/her of your intentions (to re-submit the paper or even withdraw it). If you can suggest a (realistic) likely timescale to return the paper, even better. This always helps the Editor in planning the content of near future editions of the Journal.

Collate the Criticisms

You will probably have received the referees' comments electronically. Cut and paste these into a new document. Personally, I prefer to then arrange the criticisms in the order in which they appear in the manuscript (but still noting from which referee each criticism emanates). However, it is equally as acceptable to simply arrange them reviewer by reviewer. I'm afraid that all the nice observations and positive comments made by the referees (and there probably will be some) will need to be cut out from this document. Stick to just the revisions that you need to make.

It might be recalled from the previous chapter that the reason for using two (or more) referees is not necessarily to look for concordance in their reviews but to elicit a wider range of comments and criticisms about a manuscript. As this is the case it is not too surprising if there are sometimes conflicting requirements in the referees' comments (for example, *'shorten the length of the Introduction'*: *'provide more extensive discussion of previous work describing...'*). Don't panic.

The purpose of creating a document collating all the required revisions is twofold:

- It provides a 'to do' list made up of the criticisms that you need to address to get your paper published.
- It forms the basis of the document that you will return to the Editor and referees' (along with your revised manuscript) describing exactly how and where you tackled these issues in your revised paper. This document helps to make their life much easier when assessing your revised contribution.

There are two ways to address each of the reviewers' comments. You can either agree with them (and then fix them) or you can disagree with them and then make a well argued case (as part of your response, not as part of the manuscript) about why the reviewer has got it wrong. I would suggest that you had better be on pretty firm ground before adopting the second strategy. This is not to say don't do it (reviewers do get things wrong or misunderstand things) but you will probably need to convince both reviewers and the Editor. If they have misunderstood something, then this could very well be an issue in your clarity of explanation. If you do not understand a comment from the referees, then do not hesitate to contact the Editor, who should then seek a clarification on your behalf.

In the case where there are conflicting requirements from the reviewers I bring this explicitly to the attention of the Editor and referees. Such an issue can be most clearly seen when referees' comments are collated in the order in which they occur in the manuscript. In the document describing my responses, I consider both referees' requirements and provide a short rationale concerning which comment I have given precedence to, and hence which revision has been implemented in the manuscript. This now usually becomes a problem for the Editor! However, doing nothing and ignoring both comments (hoping that they somehow cancel each other out) is not an option.

Now you have a list of the required revisions, all you have to do is fix them.

Revising Your Manuscript

Put as much effort into doing this as you put into writing the paper in the first place. Don't make it a 'minimum effort' exercise just to fix the reviewers' criticisms. You can often tell when someone has done this. For example, something is simply inserted into a paragraph to address a criticism. This makes sense in that immediate context but reading the material before and after it makes it read uncomfortably. It may introduce repetition or partially contradict something else in the manuscript. Often, if it is in the Introduction, it will not be picked up in the Discussion section (and *vice versa*). Each of the issues raised by the reviewers must be addressed individually, but the paper still needs to read as a 'whole'. If your manuscript is subject to a major re-submission, it is usually policy that the manuscript is re-reviewed in its entirety, as if it was a new paper. Making such a limited effort to revise its content will immediately be obvious.

At this point I need to say that I am extremely indebted to the reviewers of my manuscripts who over the years have unknowingly contributed the material to illustrate this section. I thank them in advance for all their hard work and their fine words that I am about to use. I only wish I could credit them more appropriately. The following are a few examples of how I have addressed referees' comments in the past. The objective is not just to address the criticisms as fully as possible and incorporate these into a revised manuscript, it is also to present the revisions and your response in such a way that it is almost impossible for the Editor not to accept the paper for publication.

The following comments were amongst those received on the initial submission of Li, Harris and Yu (2008). These are examples of simple technical comments that may be dealt with directly in the revisions:

> **Referee's Comment – p. 10**: Why is the Cohen's Kappa analysis used, given the arguments made for rejecting it as a good index of inter-rater reliability? Also one of the arguments made, concerning unreliability when the vast majority of observations fall into just one of the categories, does not apply here. What is the Kappa value for full agreement?

> **Response to referee:** A sentence explaining that coefficient Kappa ranges from 0 (no agreement) to 1 (perfect agreement) has now been included, as has a reference to the original Cohen paper. There are several categories where the Kappa value is low when the corresponding inter-rater agreement is high (e.g. 'adverse physiological states') because the majority of observations fall into one category. This example is now called out in the revised manuscript to illustrate this point.

The corresponding parts of the finally published paper were revised to read:

> Values for Kappa range between 0 and 1.00, where 0 represents complete independence between raters and 1.00 is indicative of perfect agreement (Cohen, 1960)...

> ... Cohen's Kappa becomes unreliable when the vast majority of observations fall into just one of the categories and there is also a high percentage of agreement between raters in this category (as in the category 'adverse physiological

states'—see Table 1). In such a case Cohen's Kappa will be low as the statistic is based upon expected probabilities calculated from the marginal observed totals (Huddlestone, 2003). Gwet (2002a, b) also observed that Kappa does not take in account raters' sensitivity and specificity. Gwet also observed that Kappa becomes unreliable when raters' agreement is either very small or very high.... (Li, Harris and Yu, 2008, p. 429)

Referee's Comment – p. 11, para. 2: '"Lambda" was used to calculate the proportional reduction in error'. A little more explanation is needed – how does this statistic relate to strength of association between categories – what is the potential range of values?

Response to referee: In the sub-section concerning an overview of the analysis rationale, a brief description of the function of lambda is now included, which includes a note of its potential range, and what the extremities in this range represent.

The revised manuscript read:

The lambda statistic is a measure (ranging from 0 to 1, where 1 represents certainty) of the extent that knowledge of the category of one variable improves the prediction of the other variable. Lambda (Goodman and Kruskal, 1954) has the advantage of being a directional statistic. (Li, Harris and Yu, 2008, p. 429)

Sometimes a little diplomacy is required. One of the referees brings up a good point but at the same time it is evident that they don't quite understand the objectives of the study. This can indicate poor explanation and a lack of clarity on the part of the authors but you also don't want to revise your paper to include something inappropriate or incorrect in it. The following response was made to a comment received on the initial submission of the paper by Stanton, Harris, Salmon, Demagalski, Marshall, Young, Dekker and Waldmann (2006).

Referee 3 noted that HET [Human Error Template] did not take into account multi crew aspects of the design. This issue has been noted and addressed toward the end of the Discussion section (section 5.0) in the revised paper. However, it should also be noted that HET is principally targeted at the eradication of design induced error, rather than the mitigation of design induced errors, which is one of the roles of the second crew member.

In other words, this is a good point but the referee did not completely understand the function of the new error prediction method being described in the paper. However, to incorporate the referee's suggestion the final paper was revised to read:

It should also be noted that at the moment the methodology does not encompass the potential error detection/error mitigation processes afforded when flying on a multi-crew flight deck. However, as it is a requirement that for certification purposes all aircraft are capable of operation by a single pilot without imposing undue workload, this was not regarded as a high priority item in the development of HET. (Stanton, Harris, Salmon, Demagalski, Marshall, Young, Dekker and Waldmann, 2006, p. 114)

Sometimes, comments cannot be addressed as explicitly as in the preceding cases. Nevertheless, it still needs to be made clear to the Editor and referees that their concern has been tackled appropriately. The following criticism was received on a recent submission by Bohm and Harris (in press).

Referee's Comment – Conciseness of Discussion: Discussion is too lengthy. Generally the article is too long (32 pages with figures). It would be good to shorten it a bit, especially the discussion.

Response to referee: Discussion section is now much shorter (see response to previous comment). Overall, after further editing, approximately 500 words have been removed from the manuscript (circa 10%).

These are just a few examples of responses to reviewer's criticisms. The key thing to note is that in all cases there has been an explicit and comprehensive response. The Editor and referees are also told where they can find the revisions in the manuscript (when this is possible). Some journals now require you to highlight revisions when re-submitting your paper. In the case of a manuscript that requires only minor revisions, it will be the responsibility of the Editor to check that these have been completed. Such assistance is greatly appreciated in this case. Do whatever you can to help the people responsible for producing the journal while at the same time making it almost impossible for them to refuse your re-submission.

You can now re-submit your revised paper. This is again usually done using the web-based submission system but in this case don't forget to include your document summarising the changes that you have made to the manuscript. It is also a good idea in your covering letter to include a short note of thanks to the Editor and referees for their hard work. It can take the best part of a day to read and comment on a paper, work that you benefit from and which is done *gratis* by the reviewers. If you are a PhD candidate you might benefit greatly from these comments when it comes to submitting your thesis.

As mentioned earlier, there can be several rounds of revision before the manuscript is finally accepted. The most I have ever been on the receiving end of was four re-submissions before acceptance: as a reviewer the most I have ever seen is seven. However, you will have a very good idea after just one or two rounds of revision if your manuscript is likely to be accepted. If you can't answer a major criticism from one of your referees, then it is unlikely that you will ever get the paper accepted.

Even if your paper is rejected, don't give up immediately: incorporate as many suggestions for revision as you can and then submit it to another journal (but make sure that you do the job properly and format the manuscript to the new publication's requirements). You never know your luck!

Responding to Referees' Comments: End of Chapter Checklist

Have you collated each comment into a new document to guide your revisions?	☐
Have you answered explicitly each comment and documented your response?	☐
Have you prepared a documented response to the criticisms to return with your revised manuscript?	☐
Have you been nice to the Editor and referees!	☐

Final Steps

Even when your manuscript is accepted, you are not finished yet. There are still a couple of things to do.

Copyright

At some point you will need to assign the copyright to the journal. Make sure that you have the authority to do this. If necessary, check with your employer – this can be an issue if you are a government employee. It is quite common for the UK government to insist on retaining copyright, however most journals can accommodate this. If you are the corresponding author you will also need to check with your co-contributors that you have their agreement to sign over the copyright.

Galley Proofs

After what seems like an eternity has passed you will suddenly receive an e-mail from the publisher containing a set of galley proofs for you to correct and emend, usually within 72 hours of receipt. Galley proofs (sometimes called page proofs, although these also include the final page numbers) are preliminary versions of your paper meant for your review before final printing. There may be a set of minor queries attached to them from the journal's desk editor. Typical of these are:

Please supply 3–5 keywords.

Reference 'Civil Aviation Authority 1998.' is not listed in References. Please check and confirm.

Reference 'Aeronautica Civil of the Republic of Colombia, 1995.' is not cited in the text. Please check and confirm.

However, it is your responsibility as the author to proof read the article carefully to check if any other minor corrections are required. Major revisions are not permitted at this point. Even minor textual revisions (other than typographical mistakes) are very strongly discouraged.

Proof reading is very difficult and requires a great deal of concentration. You need to read every word carefully. I never do more than one page at a time, before taking a break and doing something else. Most manuscripts can now be annotated electronically to indicate where the required emendments have been completed.

Once you have finally done this, have one last look and then press the button to save the revisions, approve the manuscript and send it back to the publishers. Congratulations: you have now finally finished your paper! In a short while it will probably appear on-line and you will be able to download a .pdf version. However, there is still nothing like getting the final, printed version in your hand. It can take some time, but I think that you will find that it is worth it. It should encourage you enough to want to write another one!

Interlude V:
A Well Kept Secret – You May Be Eligible For Some Money!

You may now find something out in this final short Interlude that could pay for the price of this book! In the long run, it could make you even more money. Unfortunately, this may only apply to British authors. This is a well-kept secret known by relatively few authors (and most of them keep it to themselves).

Whenever you inter-library loan a paper, download one or apply for permission to copy something (e.g. via 'Rightslink™' by the Copyright Clearance Center – www.copyright.com) you, as the author, are eligible to collect royalties. The UK Copyright Licence Agency (CLA) also contributes over 50% of the total monies in this fund. In the UK, this money distributed by ALCS (Authors' Licensing and Collecting Society – go to www.alcs.co.uk). ALCS also collects, on behalf of authors, all secondary royalties (i.e. not those paid to you directly by the publisher – in the case of a journal paper: none).

If you have written and published a paper you are eligible to join ALCS, which is very easy. There is a small, one-off lifetime fee (which is deducted from your first payment) and then ALCS takes a small commission from each of your payments (at the time of writing this is 9.5% for members). All you have to do is register (on-line) all your published papers (or books) that have an ISSN or an ISBN. Payments are made twice each year (September and February).

Don't get too excited at this point. It is unlikely that the royalties that you will receive from your first papers will cover more than the cost of a curry. However, if you continue to produce academic papers (and remember to register them with ALCS) the fees can begin to build up. A reasonably prolific author can collect over a thousand pounds annually after a few years. It really costs you nothing to join, and remember – this is actually money that you have earned!

Some time ago I mentioned that it was a good idea to get your keywords, Title and Abstract right. Now can you see why?

Chapter 11
A Few Final Thoughts

If you have got this far (successfully) well done! Seeing your research paper published is a great feeling. I sincerely hope that it encourages you enough to want to write more.

You might like to follow the recipe in this book for the next one or two manuscripts but then I strongly suggest that you throw it away (or give it to someone else, or sell it on eBay to get some of your money back). If you want to keep anything, rip out the end of chapter check lists, but please get rid of the rest of it. It is important that you develop your own style and your own way of doing things. This book contains just one way of producing workmanlike Human Factors papers. I have found that it works for me and that other people who have tried this approach have also found it works. However, the world would be a dull place if all papers were produced to the same template. I hope that you eventually find a better way that works for you.

One of the things that you can best do to accelerate your critical reading and writing skills is review the work of others, either for conferences or for journals. If you get the chance, do it. The whole scientific journal enterprise depends heavily on the goodwill of Editors and referees who donate their time and expertise freely. Hopefully, at some point you will be asked to review manuscripts. Lots of people have given their time (and will give their time) to review your manuscripts. Please return the favour. There are those authors known to all Editors who never hesitate to submit their papers to journals but who are always too busy to review manuscripts.

Never be afraid to try and make your manuscripts interesting. John Reith summarised the mission of the BBC in three key words: to *educate, inform* and *entertain.* These are pretty good goals when you write scientific papers (even if they are usually a little short on entertainment). A reader's reaction to any piece of writing is as much emotional as it is rational. If you can write manuscripts that engage the reader, this is to be admired. Writing itself is also to be enjoyed. It should not be a chore. And one of the best ways for it to remain interesting is to keep trying something different. Remember this when you are an Editor.

Happy writing!

References

Albery, I. P. & Guppy, A. (1995). The interactionist nature of drinking and driving: A structural model. *Ergonomics, 38,* 1805–1818.

Bell, H. H. & Lyon, D. R. (2000). Using observer ratings to assess Situational Awareness. In, M. R. Endsley & D. J. Garland (Eds) *Situation Awareness Analysis and Measurement* (pp. 129–146). New Jersey: Laurence Earlbaum Associates.

Berger, D. E. & Snortum, J. R. (1986). A structural model of drinking and driving: Alcohol consumption, social norms and moral commitments. *Criminology, 24,* 139–153.

Bohm, J. & Harris, D. (2010). Risk perception and risk-taking behavior of construction site dumper drivers. *International Journal of Occupational Safety and Ergonomics, 16,* 55–67.

Bohm, J. & Harris, D. (in press). Hazard awareness of construction site dumper drivers. *International Journal of Occupational Safety and Ergonomics.*

Demagalski, J. M., Harris, D. & Gautrey, J. E. (2002). Flight control using only engine thrust: development of an emergency display system. *Human Factors and Aerospace Safety, 2,* 173–192.

Dennis, K. A. & Harris, D. (1998). Computer-based simulation as an adjunct to ab initio flight training. *International Journal of Aviation Psychology, 8,* 261–276.

Drury, C. G., Prabhu, P. & Gramopadhye, A. (1990). Task analysis of aircraft inspection activities: methods and findings. In, *Proceedings of Human Factors Society 34th Annual Meeting* (pp. 1181–1185). Santa Monica, CA: Human Factors and Ergonomics Society.

Dunbar, J. A., Penttila, A. & Pikkarainen, J. (1987). Drinking and driving: Success of random breath testing in Finland. *British Medical Journal, 295,* 101–103.

Ebbatson, M., Harris, D., Huddlestone, J. & Sears, R. (2008). Combining control input with flight path data to evaluate pilot performance in transport aircraft. *Aviation Space and Environmental Medicine, 79,* 1061–1064.

Ebbatson, M., Huddlestone, J., Harris, D. & Sears, R. (2006). The application of frequency analysis based performance measures as an adjunct to flight path derived measures of pilot performance. *Human Factors and Aerospace Safety 6,* 383–394.

Ezzy, D. (2002). *Qualitative analysis: practice and innovation.* London, UK: Routledge.

Gaur, D. (2005). Human Factors Analysis and Classification System applied to civil aircraft accidents in India. *Aviation, Space and Environmental Medicine, 76*, 501–505.

Gibson, J. C. (1995). *The definition, understanding and design of aircraft handling qualities (Rep. No. LUT LR–756)*. Delft, The Netherlands: Delft University of Technology.

Glaser, B. & Strauss, A. (1967). *The discovery of Grounded Theory*. Chicago, IL: Aldine.

Goodman, L. & Kruskal, W. H. (1954). Measures of association for cross-classifications. *Journal of the American Statistical Association, 49*, 732–764.

Guppy, A. (1988). Factors associated with drink-driving in a sample of English males. In, J. A. Rothengatter & R. A. de Bruin (Eds) *Road user behavior: theory and research* (pp. 375–380). Assen, The Netherlands: Van Gorcum.

Gwet, K. (2002a). Kappa statistics is not satisfactory for assessing the extent of agreement between raters. *Statistical methods for inter-rater reliability assessment, 1*, 1–6. Retrieved 24 November 2006 from www.stataxis.com/files/articles/inter_rater_reliability_ dependency.pdf

Gwet, K. (2002b). Inter rater reliability: Dependency on trait prevalence and marginal homogeneity. *Statistical methods for inter-rater reliability assessment, 1*, 1–9. Retrieved 24 November 2006, from www.stataxis.com/files/articles/inter_rater_reliability_ dependency.pdf.

Harris, D. (2000). The measurement of pilot opinion when assessing aircraft handling qualities. *Measurement and Control, 33*, 239–243.

Harris, D. (2002). Drinking and flying: causes, effects and the development of effective countermeasures. *Human Factors and Aerospace Safety, 2*, 297–318.

Harris, D. (2005). Drinking and flying: causes, effects and the development of effective countermeasures. In, D. Harris & H. C. Muir (Eds) *Contemporary Issues in Human Factors and Aviation Safety* (pp. 199–219). Aldershot, UK: Ashgate.

Harris, D. & Maxwell, E. (2001). Some considerations for the development of effective countermeasures to aircrew use of alcohol while flying. *International Journal of Aviation Psychology, 11*, 237–252.

Harris, D. & Rees, D. (1993). The use of fly-by-wire controls in a primary training aircraft: implications for pilot skill development. *Aerogram, 7*, 3–4.

Harris, D., Chan-Pensley, J. & McGarry, S. (2005). The development of a multidimensional scale to evaluate motor vehicle dynamic qualities. *Ergonomics, 48*, 964–982.

Harris, D., Gautrey, J., Payne, K. & Bailey, R. (2000). The Cranfield Aircraft Handling Qualities Rating Scale: a multidimensional approach to the assessment of aircraft handling qualities. *The Aeronautical Journal, 104 (March)*, 191–198.

Harris, D., Payne, K. & Gautrey, J. (1999). A multidimensional scale to assess aircraft handling qualities. In, D. Harris (Ed.) *Engineering Psychology and Cognitive Ergonomics (Volume Three – Transportation, Medical Ergonomics and Training Systems, pp. 277–285).* Ashgate, UK: Aldershot.

Hart, S. G. & Staveland, L. E. (1988). Development of NASA-TLX (Task Load Index): results of empirical and theoretical research. In, P. A. Hancock & N. Meshkati (Eds) *Human Mental Workload* (pp. 139–183). Amsterdam, The Netherlands: North-Holland.

Hays, W. L. (1988). *Statistics (4th Edition).* New York, NY: Holt, Reinhardt and Winston.

Homel, R., Carseldine, D. & Kearns, I. (1988). Drink-driving countermeasures in Australia. *Alcohol, Drugs and Driving, 4,* 113–144.

Hubbard, D. C. (1987). Inadequacy of root mean square error as a performance measure. In, R. S. Jensen (Ed.) *Proceedings of the Fourth International Symposium on Aviation Psychology* (pp. 698–704). Columbus, OH: Ohio State University.

Huddlestone J. A. (2003). *An evaluation of the training effectiveness of a low-fidelity, multi-player simulator for air combat training* [Ph.D. thesis]. Bedford, UK: College of Aeronautics, Cranfield University: 2003.

Huddlestone, J. & Harris, D. (2007). Using Grounded Theory techniques to develop models of aviation student performance. *Human Factors and Aerospace Safety, 6,* 357–368.

Jarvis, S. R. & Harris, D. (2007). Looking for an accident: Glider pilots' visual management and potentially dangerous final turns. *Aviation Space and Environmental Medicine, 78,* 597–600.

Jarvis, S. R. & Harris, D. (2008). Investigation into accident initiation events by flight phase, for highly inexperienced glider pilots. *International Journal of Applied Aviation Studies, 8,* 211–224.

Jarvis, S. R. & Harris, D. (2010). Development of a bespoke human factors taxonomy for gliding accident analysis and its revelations about highly inexperienced UK glider pilots. *Ergonomics, 53,* 294–303.

Jorna, P. G. A. M. & Hoogeboom, P. J. (2004). Evaluating the flight deck. In D. Harris (Ed.), *Human factors for civil flight deck design* (pp. 235–274). Aldershot, UK: Ashgate.

Kinkade, P. T. & Leone, L. M. C. (1992). The effects of 'tough' drunk-driving laws on policing: A case study. *Crime and Delinquency, 38,* 239–257.

Landis, J. R. & Koch, G. G. (1977). The measurement of observer agreement for categorical data. *Biometrics, 33,* 159–174.

Li, W-C. & Harris, D. (2006). Pilot error and its relationship with higher organizational levels: HFACS analysis of 523 accidents. *Aviation, Space and Environmental Medicine, 77,* 1056–1061.

Li, W-C., Harris, D. & Yu, C-S. (2008). Routes to failure: analysis of 41 civil aviation accidents from the Republic of China using the Human Factors Analysis and Classification System. *Accident Analysis and Prevention, 40,* 426–434.

Lock, M. W. B. & Strutt, J. E. (1985). *Reliability of in-service inspection of transport aircraft structures (CAA Paper 85013).* London, UK: Civil Aviation Authority (HMSO).

Mäkinen, T. (1988). Enforcement studies in Finland. In, J. A. Rothengatter & R. A. de Bruin (Eds) Road user behavior: Theory and research (pp. 584–588). Assen, The Netherlands: Van Gorcum.

Martinussen, M. & Hunter, D. (2010). *Aviation Psychology and Human Factors.* Boca Raton FL: CRC Press.

Maxwell, E., & Harris, D. (1999). Drinking and flying: A structural model. *Aviation, Space and Environmental Medicine, 70,* 117–123.

McKnight, A. J. & Voas, R. B. (1991). The effect of license suspension upon DWI recidivism. *Alcohol, Drugs and Driving, 7,* 43–54.

Norström, T. (1978). Drunken driving: a tentative causal model. *Scandinavian Studies in Criminology, 6,* 252–283.

Norström, T. (1981). Drunken driving: a causal model. In, L. Goldberg (Ed.) *Alcohol, drugs and traffic safety: proceedings of the 8th International Conference on Alcohol, Drugs and Traffic Safety* (pp. 1215–29). Stockholm, Sweden: Almqvist and Wiksel.

Ostberg, O. (1980). Risk perceptions and work behavior in forestry: implications for accident prevention policy. *Accident Analysis and Prevention, 12,* 189–200.

Payne, K. & Harris, D. (2000). The development of a multi-dimensional aircraft handling qualities rating scale. *International Journal of Aviation Psychology, 10,* 343–362.

Psymouli, A., Harris, D. & Irving, P. (2005). The inspection of aircraft composite structures: a Signal Detection Theory-based Framework. *Human Factors and Aerospace Safety, 5,* 91–108.

Rees, D. J. & Harris, D. (1995). Effectiveness of *ab initio* flight training using either linked or unlinked primary-axis flight controls. *International Journal of Aviation Psychology, 5,* 291–304.

Ross, H. L. (1988). Deterrence-based policies in Britain, Canada and Australia. In, M. D. Lawrence J. R. Snortum & F. E. Zimring (Eds) *Social control of the drinking driver* (pp. 64–78). Chicago, IL: University of Chicago Press.

Ross, L. R. & Ross, S. M. (1992) Professional pilots' evaluation of the extent causes and reduction of alcohol use in aviation. *Aviation, Space and Environmental Medicine, 63,* 805–808.

Ross, S. M. & Ross, L. R. (1995). Professional pilots' views of alcohol use in aviation and the effectiveness of employee assistance programs. *International Journal of Aviation Psychology, 5,* 199–213.

Sloane, H. R. & Cooper, C. L. (1984). Health-related lifestyle habits in commercial airline pilots. *British Journal of Aviation Medicine, 2,* 32–41.

Stanton, N. A. & Young, M. S. (1999a). Utility analysis in cognitive ergonomics. In, D. Harris (Ed.) *Engineering Psychology and Cognitive Ergonomics* (Volume Four, pp. 411–418). Aldershot, UK: Ashgate.

Stanton, N. A. & Young, M. S. (1999b). *A guide to methodology in ergonomics.* London: Taylor and Francis.

Stanton, N. A., Harris, D., Salmon, P., Demagalski, J. M., Marshall, A., Young, M. S., Dekker, S. W. A. & Waldmann, T. (2006). Predicting design induced pilot error using HET (Human Error Template) – a new formal human error identification method for flight decks. *The Aeronautical Journal, 110 (February)*, 107–115.

Strauss, A. & Corbin, J (1990). *Basics of qualitative research: grounded theory procedures and techniques.* London, UK: Sage.

van Doorn, R. R. A. & de Voogt, A. J. (2007). Glider accidents: an analysis of 143 cases, 2001–2005. *Aviation, Space and Environmental Medicine, 78,* 26–28.

Vingilis, E. R. & Salutin, L. (1980). A prevention programme for drinking and driving. *Accident Analysis and Prevention, 12,* 267–274.

Waag W. L. & Houk M. R. (1994). Tools for assessing Situational Awareness in an operational fighter environment. *Aviation, Space and Environmental Medicine, 64 (5, Suppl)* A13–A19.

Weyman, A. K. & Clarke, D. D. (2000). Investigating the influence of organisational role on perceptions of risk in deep coal mines. *Journal of Applied Psychology, 88,* 404–412.

Wheeler, G. R. & Hissong, R. V. (1988). Effects of criminal sanctions on drunk drivers: Beyond incarceration. *Crime and Delinquency, 34,* 29–42.

Widders, R. & Harris, D. (1997). Pilots' knowledge of the relationship between alcohol consumption and levels of blood alcohol concentration. *Aviation, Space and Environmental Medicine, 68,* 531–537.

Wiegmann, D. A. & Shappell, S. A. (2003). *A human error approach to aviation accident analysis: the Human Factors Analysis and Classification System.* Aldershot, UK: Ashgate.

Appendices

Preface to Appendices

I have chosen these four particular papers for inclusion in the Appendices not because they are especially outstanding but because they each exemplify a different type of Human Factors paper. Each paper also had its own particular challenges. These papers also serve to illustrate many of the points and principles described in the book.

The first two papers (Harris and Maxwell, 2001 and, Harris, McGarry and Chan-Pensley, 2005) are each based upon large scale survey research. Both papers required multi-stage analysis using multivariate statistical techniques. Harris, McGarry and Chan-Pensley (2005) was also a multi-stage, multi-study paper. Huddlestone and Harris (2007) is a paper using the analysis of qualitative data using a grounded theory-based approach. Finally, Demagalski, Harris and Gautrey (2002) is a study using an experimental methodology, but conducted in the ecologically-valid context of a flight simulator. The papers are also published in three different journals, which gives and appreciation of the slightly differing requirements in publications.

As a short introduction to each paper I have provided a snapshot of the background of the research; a synopsis of the particular challenges each paper posed; and the 'Big Message' underlying each manuscript, along with the bullet points that guided the story. I hope that these short summaries give some insights into the way each paper was constructed.

Appendix 1:
Harris and Maxwell (2001)

Harris, D. & Maxwell, E. (2001). Some considerations for the development of effective countermeasures to aircrew use of alcohol while flying. *International Journal of Aviation Psychology, 11,* 237–252.

Commentary

Just over a decade ago, the UK CAA introduced a specific blood alcohol limit above which it was prohibited to act as a pilot in an aircraft. This limit was very low (just one-quarter of that prescribed in UK law for drinking and driving). This paper was one of the last in a series that looked at the effects of low blood alcohol levels on pilot performance (see Harris, 2002: 2005). It was a direct follow on paper from Widders and Harris (1997) which found that pilots may either inadvertently transgress the limit as a result of not having the required knowledge about the rate at which alcohol is eliminated from the body, or the fact that they did not agree with such a low limit (so were prepared to ignore it). Furthermore, despite a great deal of research in the road safety domain, little of the knowledge gained in this area had been applied to the deterrence of drinking and flying. This was what this paper was all about.

The challenges in writing this paper were twofold. Firstly, we were writing for an aviation audience, so they had to be appraised about the most important papers from the large body of research emanating from the road safety domain. Secondly, both the treatment of the data (to form the dependent and independent variables) and its subsequent analysis were slightly convoluted which required a careful structuring and on-going interpretation for the reader. The tables were also quite complex. In general, there was quite a lot going on in the paper all of which had to be summarised within about 6,500 words. The challenges the material posed were ideal for illustrating many key aspects of the principles described in this book.

The 'Big Message'

You need to use the right countermeasures to reduce the likelihood of drinking and flying otherwise all your efforts will go to waste – different people drink and fly for different reasons.

The Story

- Drinking and flying is undesirable – new regulations are on the horizon (at the time of writing).
- A great deal is known from the development of drinking and driving countermeasures but these have never been applied to aviation.
- Effective countermeasures depend upon the reasons for the underlying offending behaviour.
- A survey of 400+ pilots identified four theoretically recognisable factors for drinking and flying countermeasures: Education; Enforcement; Counselling; Sanctions.
- Different countermeasures were rated as being more likely to be effective in different groups of pilot (defined by licence type and drink/flying category).
- The results support specific deterrence theory (this was the main scientific 'hook').

Once you are aware of this background, see if you can identify how the 'Big Message' and the story underpin the manuscript.

THE INTERNATIONAL JOURNAL OF AVIATION PSYCHOLOGY, 11(3), 237–252)
Copyright © 2001, Lawrence Erlbaum Associates, Inc.

FORMAL PAPERS

Some Considerations for the Development of Effective Countermeasures to Aircrew use of Alcohol While Flying

Don Harris and Emma Maxwell
College of Aeronautics
Cranfield University, Cranfield, Bedford, England

A revision to the Joint Airworthiness Authorities' Operations Regulations now imposes a maximum Blood Alcohol Concentration limit of just 0.02% on U.K. pilots. Using a postal survey, opinions were elicited from 472 private and professional pilots concerning the effectiveness of various countermeasures to reduce the likelihood of drinking and flying. Punitive sanctions and tougher enforcement of the regulations were regarded as the most effective countermeasures, although offenders and professional pilots thought these actions less effective than private pilots and nonoffenders. The results are discussed with respect to producing effective countermeasures specifically targeted at high-risk groups.

The operation of an aircraft is specifically prohibited when under the influence of alcohol. No amount of alcohol whatsoever is sanctioned within the blood of on-duty aircrew. In 1985, for legal purposes, the U.S. Federal Aviation Administration (FAA) established a specific upper blood alcohol concentration (BAC) of 40mg per 100ml of blood (also expressed as 0.04%) above which it was expressly prohibited to act as a crew member of a civil aircraft. In 1995 this rule was amended to incorporate a provision for evidentiary breath testing. In the case of a pilot pro-

Requests for reprints should be sent to Don Harris, College of Aeronautics, Cranfield University, Cranfield, Bedford MK43 0AL, England. E-mail: d.harris@cranfield.ac.uk

ducing a positive test for alcohol with a BAC of between 0.02% and 0.039%, that crew member now cannot fly until a repeat test indicates that their BAC is below 0.02% or at least 8 hrs have elapsed from taking the initial test. FAA regulations also include provision for the suspension or revocation of an offender's license. Until April 1998, Article 57 of the U.K. Air Navigation Order (No. 2; Department of Transport, 1995) merely stated that "the limit of drinking or drug taking is any extent at which the capacity to act as a crew member would be impaired." This regulation was subsequently amended to incorporate a revision to the European Joint Aviation Authorities operations regulations (JAR OPS) that specifically prohibits a pilot to act as a crew member with a BAC of greater than 0.02%. For the average 80kg man, this corresponds to drinking just over 1/2 pint (254ml) of normal strength beer.

Widders and Harris (1997) found that up to 24% of U.K. pilots could not determine when their BAC was likely to fall below the 0.02% level after drinking and may therefore inadvertently infringe the regulation. A further large proportion of pilots felt that they were safe to fly before their BAC had dropped below 0.02%, indicating a willingness to infringe the regulations. These two hypothetical groups of pilots, which comprised almost 50% of the sample, were termed *inadvertent drink flyers* and *non-believers*, respectively. The inability of pilots to estimate their BAC after drinking has also been suggested as a contributory factor to drinking and flying by other researchers (e.g., Ross & Ross, 1992). This phenomenon is not unique to pilots. A study of car drivers observed that a considerable proportion was not likely to drive during the absorption phase but would not leave long enough for their body to eliminate the alcohol, hence they were at risk of driving while intoxicated during the elimination phase (Bierness, 1987).

The study of drinking and driving has a great deal to offer in the development of effective measures to discourage drinking and flying, however, little of this literature has been referred to in this context. Vingilis and Salutin (1980) described a tripartite model for the deterrence of drink-driving behavior. Primary level interventions were essentially concerned with educating drivers about the effects of alcohol on performance. Countermeasures at the secondary level were concerned with enforcement of the regulations (i.e. increasing the likelihood of apprehension of offenders). The tertiary level in the model aimed to reduce the chances of recidivism through either punitive sanctions (aimed at suppressing offending behavior) or through counseling and rehabilitation (targeted at eliminating the root cause of offending behavior).

Educating pilots about BAC decay rates would only be an effective drink-flying countermeasure for those pilots falling into the "inadvertent drink-flyer" group. However, this group comprised only 16% of offenders (Widders & Harris, 1997). The majority of potential infringers of the 0.02% rule were non-believers. Nörstrom (1978, 1981), Berger and Snortum (1986) and Albery and Guppy (1995) all identified a lack of a moral attachment to the law as a key determinant of drink-driving behavior. Thus, effective countermeasures for non-believers are likely to be different to those of the inadvertent drink-flyer group as the root cause of the offending behavior is different. Effective remedial actions can only be specified with regard to the cause of the offending behavior, not the behavior itself. Thus, for non-believers the emphasis in any intervention strategies should be placed on the latter two aspects of the tripartite model whereas inadvertent drink-flyers should respond best to primary level interventions.

The effect of enforcement (secondary level intervention) on suppressing drink-driving behavior is well understood. Increasing the perceived likelihood of arrest acts as an effective deterrent (Guppy, 1988). Increasing the actual likelihood of arrest when drink-driving also suppresses this behavior. The introduction of large-scale random breath testing (RBT) in Finland and Australia led to beneficial effects on the suppression of offending behavior (Dunbar, Penttila, & Pikkarainen, 1987; Homel, Carseldine & Kearns, 1988). In Finland a 58% reduction in the level of drinking and driving was observed after the introduction of RBT; analysis of Australian data showed a 42% reduction in the number of alcohol related road traffic accidents.

Vingilis and Salutin (1980) argued tertiary level interventions were essentially designed to prevent recidivism not to deter offending behavior. To progress to the tertiary stage the offender must have passed through the previous stages. Classical deterrence theory, however, would suggest that sanctions could potentially operate at all stages in the tripartite model. *General* deterrence acts on all members of the population and should deter potential offenders from committing the illegal act; *specific* deterrence is concerned with the effect of punitive sanctions on offenders to reduce the likelihood of recidivism (see Ross, 1984).

The effectiveness of *general* deterrence as a drinking and driving countermeasure has been difficult to establish. Mäkinen (1988) reported that from a review of over 200 enforcement studies, severe punishment had no demonstrable effect on improving drivers' behavior. Ross (1988) supported this finding, suggesting that increasing the likelihood of apprehension on a drink-driving occasion (Stage 2 of the model) was a more effective deterrent. However, later research has suggested that widely advertised punitive sanctions could also act as a deterrent to initial offending (Kinkade & Leone, 1992). The implementation of severe sanctions for drink-driving in California reduced the arrest rate by approximately 12%. The effectiveness of punitive sanctions as a *specific* deterrent (i.e. applied to offenders to deter re-offending) would also seem to be limited. Wheeler and Hissong (1988) observed no difference on the likelihood recidivism with respect to the severity of the sanction imposed (probation, fine or imprisonment).

In a meta-analysis of 215 studies evaluating rehabilitation programs for drinking and driving offenders (the alternative approach to sanctions at the tertiary level of the tripartite model), Wells-Parker, Bangert-Drowns, McMillen and Williams (1995) concluded this form of remedial action resulted in a 7% to 9% reduction in re-offending. Reviews of research comparing the effect of sanctions and rehabilitation programs also concluded that in the longer term, alcohol rehabilitation programs were of greater benefit than the suspension of drivers' licenses (McKnight & Voas, 1991; Peck, 1991).

Maxwell and Harris (1999) used a structural equation modeling approach to predict drink-flying offending behavior. Offending behavior was best described as a combination of personal factors (e.g., indoctrination into the culture of professional aviation) and situational factors (e.g., job-related stresses). Professional pilots holdinganairline transport pilot's licence (ATPL) were found to have the highest mean weekly consumption of alcohol. Sloane and Cooper (1984) also reported that 12% of professional pilots drank as a means of coping with personal or work-related stress. An opinion-survey of U.S. professional pilots suggested that the provision of employee assistance programs (a tertiary level countermeasure) would be the most effective approach to reducing the likelihood of drinking and flying (Ross & Ross, 1995). This was followed by (in decreasing order of effectiveness) education of pilots on the effects of alcohol and strengthening sanctions. Rehabilitation programs for pilots with alcohol problems have shown considerable success in the U.S., with positive results in up to 85% of cases (Russel & Davis, 1995; Flynn, Sturges, Swarsen, & Kohn, 1993).

Effective countermeasures can only be stipulated with respect to the reasons for the offending behavior. Thereasons whyinadvertent drink-flyers and non-believers may infringe the regulations are quite different. inadvertent drink-flyers may offend through a lack of knowledge about the rate at which alcohol is eliminated from the body; non-believers may offend as a result of a lack of moral commitment to the legislation. Professional pilots may also drink as a result of stress from organizational pressures. This study examines the relative effectiveness of the components of the tripartite model for the deterrence of drinking and flying with respect to the type of offender and license category of pilot. The objective is to identify effective countermeasures to drinking and flying, with specific reference to the 0.02% BAC rule specified in the 1998 revision of JAR OPS.

METHOD

Sample

A random sample of 1,000 holders of pilot's licenses, from all categories, was obtained from the U.K. Civil Aviation Authority (CAA). To preserve the anonymity of respondents and comply with the terms and provisions of the U.K.'s 1984 Data Protection Act, names and addresses of the sample were not provided to the researchers.Questionnaires were placed in blank envelopes with a FREEPOST return envelope for the completed survey. These were delivered to the CAA flight crew licensing department who distributed them on behalf of the researchers. Completed questionnaires were returned directly to Cranfield University, thereby maintaining the independence of the research from the regulatory authority and assuring the anonymity of the respondents.

Questionnaire Content

Respondents were required to provide brief demographic data; for example sex, age, licenses held, and total number of flying hours. Respondents were also asked for their weight, as this was necessary for the subsequent calculation of BAC.

In the next section of the survey instrument, respondents were then required to estimate (in hours) how long they thought that it would take for their BAC to drop below 0.02% in nine situations: having consumed 2, 4, or 6 pints (568 ml glasses) of beer or lager; having consumed 2, 4, or 6 measures (25 ml) of their preferred spirit; and having consumed 2, 4, or 6 standard glasses (125 ml) of wine. Respondents were also requested to state how long they would leave after drinking alcohol before they regarded themselves as being safe to fly, *irrespective* of the CAA's 0.02% BAC regulation. This was done for the nine drinking situations described previously. These items followed the format previously employed by Widders and Harris (1997).

The following section of the questionnaire required respondents to rate the effectiveness of various educational or enforcement approaches for reducing the likelihood of drinking and flying. These items were in the form of a 7-point Likert-type scale with responses ranging from 1 (*ineffective*)to7 (*most effective*).

Six items referring to the effectiveness of the various approaches were derived from items employed by Ross and Ross (1995). These referred to the provision of counseling programs to help pilots with personal problems; educating pilots about the effects of alcohol on flying performance; educating pilots about stress and relaxation strategies; the potential benefits of advising pilots on various health topics related to nutrition, exercise, and so on; appraising pilots about how alcohol effects the body; and the effectiveness of providing mandatory education programs about the effects of alcohol. Seven further items, derived from the drinking and driving literature, examined aspects of enforcement or deterrence. These included questions seeking opinions on the effectiveness of implementing tougher regulations and sanctions; increasing the level of enforcement of the drinking and flying regulations; implementing random breath checks; mandatory rehabilitation programs for alcohol abusers; revoking pilot's licenses of alcohol abusers; revoking pilot's licenses of pilots apprehended for drink-driving; and the introduction of fines for drinking and flying.

RESULTS

Sample

Four hundred and seventy-two completed survey instruments were returned. Nineteen further questionnaires were returned as being undeliverable, resulting in a final response rate of 48.1%. This ensured a probability of 0.95 that the sample obtained was within ± 0.1 *SD* of the true population mean (Hays, 1988).

Table 1 describes the composition of the sample broken down by aircraft and license category. Comparative U.K. data for the number of pilots in each category are also included. It should be noted that as CAA licensing records are held on a license-by-license and not a pilot-by-pilot basis, a pilot will appear on the database on several occasions if they hold more than one license. This caveat should be borne in mind when interpreting these data.

Of the final sample, 448 respondents were men and 22 were women. Data were missing in two cases. The mean age of the sample was 42.76 years old (with an *SD* of 11.48). Mean flight experience was 4,069.49 hrs, with a slight skew (1.28) toward the sample being composed of more experienced pilots. The corresponding standard deviation was 5,244.13 hr.

Treatment of Data—Categorization of Respondents

Respondents were classified into one of the three hypothetical offender categories identified by Widders and Harris (1997), for example, Non-Drink Flyers; Non-Believers; and Inadvertent Drink-Flyers. This categorization process was undertaken in the following way.

In each of the drinking situations described the respondent's theoretical peak BAC was calculated from the formula proposed by Snortum and Berger (1989), where: BAC = 3.75 × (Number of standard drinks/Body weight in Kg × 2.2). One standard drink was defined as the equivalent of 0.5 U.S. fluid ounces (11.67g) of alcohol. For the purposes of calculation, 1 pint (568 ml) of normal strength beer (3.85% alcohol by volume) was estimated to contain 16g of alcohol; a single measure of spirits (50% ABV) was deemed to contain 8 grams of alcohol and a glass of wine (10% ABV) was estimated to contain 10g. This formula has been found to re-

TABLE 1
Composition of the Obtained Sample Broken Down by Aircraft Type and Licence Category[a]

Licence Type	Fixed Wing		Rotary Wing		Sample Totla		U.K. Pilot's Licences	
	n	%	n	%	n	%	n	%
PPL	231	48.94	8	1.69	239	50.63	27,524	68.71
CPL	44	9.32	1	0.21	45	9.53	4,126	10.30
ATPL	159	33.69	29	6.15	188	39.84	8,410	20.99
Total	434	91.95	38	8.05	472	100.00	40,060	100.00

Note. PPL = private pilot's licence; CPL = commercial pilot's licence; ATPL = airline transport pilot's licence.

[a]Comparative U.K. data for the number of pilots' licences in each category are also included.

sult in a relatively high BAC estimate (see Widders & Harris, 1997 and Widders, 1994 for reviews of the relative merits of the formulae that can be utilized to estimate BAC). It was considered that a formula resulting in a high BAC estimate would be most appropriate when undertaking research with implications for flight safety.

After calculating a respondent's peak BAC in each of the nine drinking scenarios it was assumed alcohol would be eliminated at 0.015% per hour, a typical figure used by many researchers (e.g., Drew, Colquhoun, & Long, 1959; Balfour, 1988; Ross & Ross, 1988; Ross & Ross 1990; Mertens, Ross, & Mundt, 1991).

On the basis of these calculations, and on the answers to the items requiring respondents to state how long they would leave after drinking alcohol before they regarded themselves as being safe to fly irrespective of the CAA's 0.02% BAC regulation, respondents were categorized into one of the three drink-flying offender groups. Those with a BAC in any of the drinking scenarios estimated to be over 0.02% at the time, at which they thought that their BAC would be below this figure, were included in the inadvertent drink-flyers group. Pilots who thought that they were safe to fly before they estimated their BA Chad dropped below 0.02%, irrespective of the accuracy with which they could estimate their BAC, were categorized as non-believers. Respondents who fell into both offender categories were entered into the non-believers group. Respondents falling into neither category were designated non-drink flyers. The results from this analysis are described in Table 2.

Using the results for the whole sample as a reference distribution for subsequent chi-square "goodness-of-fit" analyzes, it was observed that private pilot's licence holders were overrepresented in the non-drink flyers category ($x^2 = 8.96$; $df = 2$; $p < 0.05$). Commercial pilot's licence (CPL) holders were not over represented in any of the offender categories ($x^2 = 0.14$; $df = 2$; $p > 0.05$). Holders of ATPLs, however, were over represented in the non-believers category ($x^2 = 7.65$; $df = 2$; $p < 0.05$).

TABLE 2
Distribution of Respondents in Each Drink-Flying Category Broken Down by License Type

	Non-Drink Flyers		*Non-Believers*		*Inadvertent Drink Flyers*		*Row Total*	
Licence Type	n	%	n	%	n	%	n	%
PPL	155	64.9	55	23.0	29	12.1	239	50.6
CPL	24	53.3	15	33.3	6	13.3	45	9.5
ATPL	90	47.9	77	41.0	21	11.2	188	39.8
Total	269	57.0	147	31.1	56	11.9	472	

Note. In some cells, the number of observations is quite small, so in these instances, the results should be interpreted with some caution. PPL = private pilot's licence; CPL = commercial pilot's licence; ATPL = airline transport pilot's licence.

Treatment of Education and Enforcement Survey Data

The items soliciting respondent's opinions on the effectiveness of the potential strategies that could be employed to reduce the occurrence of drinking and flying were subject to a principle components analysis (PCA) to produce a parsimonious data set for subsequent analysis. These results are described in Tables 3 through 5.

The Kaiser-Meyer-Olkin measure of sampling adequacy for these data was 0.75. Measures of sampling adequacy items varied between 0.68 and 0.86, suggesting these data were suitable for PCA.

Using Kaiser's rule, the PCA resulted in four factors being identified. The loadings for each variable on each factor (after varimax rotation) are shown in Table 4. Factor statistics and their labels (assigned after interpretation to reflect their constituent components) are given in Table 5.

Four clearly distinct factors pertaining to different aspects of potential drinking and flying countermeasures were identified in the PCA. Even though one of the enforcement variables (mandatory rehabilitation programs) was included on factor three (counseling and rehabilitation), the variable was clearly related to the other variables included on that factor. Scale scores for each respondent were constructed from the factors extracted by summing the raw scores for the high-loading variables on each factor and dividing the total by the number of components in the scale. Scale scores were preferred to stan-

TABLE 3

Mean, Standard Deviation, and Valid Number of Responses for the Whole Sample, for Each of the Education and Enforcement Survey Items Assessing the Effectiveness of the Various Drink-Flying Countermeasures Proposed[a]

Survey Item	M	SD	n
Counseling for pilots with personal or family problems	3.29	1.75	454
Education about alcohol's effects on flying performance	5.54	1.56	458
Education in stress reduction strategies	3.85	1.71	458
Education on health topics	4.34	1.67	458
Education about how alcohol affects the body	5.25	1.60	458
Mandatory education programs	3.67	1.87	458
Tougher drink-flying regulations	4.72	1.79	457
Increased enforcement of regulations	5.23	1.52	457
Random breath testing	5.62	1.66	458
Mandatory rehabilitation programs	4.43	1.91	458
Revoke pilot's licences of alcohol abusers	5.99	1.48	460
Revoke pilot's licences of drink–drive offenders	4.59	2.19	455
Introduction of fines for drinking and flying offenders	4.55	1.93	458

[a]High scores are indicative of a more effective approach to the elimination of drink-flying behavior.

TABLE 4

Factor Loadings After Varimax Rotation (Correlations With the Weighted Linear Combination) for Each of the Education and Enforcement Survey Items Assessing the Effectiveness of the Various Drink-Flying Countermeasures Proposed

	Factor Loading			
Variable	*1*	*2*	*3*	*4*
Education about alcohol's effects on flying	0.89	0.01	-0.05	0.07
Education about how alcohol affects the body	0.88	0.04	0.07	0.09
Education on health topics	0.64	-0.03	0.47	0.03
Mandatory education programs	0.49	0.37	0.27	0.05
Tougher drink-flying regulations	0.07	0.90	0.05	0.13
Increased enforcement of regulations	0.15	0.80	0.03	0.18
Random breath testing	-0.09	0.71	0.11	0.21
Counseling for pilots with personal or family problems	0.01	0.09	0.73	-0.12
Education in stress-reduction strategies	0.30	0.02	0.71	0.04
Mandatory rehabilitation programs	0.01	0.10	0.68	0.32
Revoke pilot's licenses of drink–drive offenders	-0.07	0.14	0.02	0.78
Introduction of fines for drinking and flying	0.20	0.11	0.12	0.71
Revoke pilot's licenses of alcohol abusers	0.10	0.39	-0.04	0.65

TABLE 5

Descriptive Labels Assigned to Each of Factors Extracted From the Principle Components Analysis and Factor Summary Statistics (Pre- and Post Varimax Rotation)

Factor	*Label*	*Eigenvalue*	*% Variance*	*Post-Rotated Eigenvalue*	*% Variance (Post Rotation)*
1	Education	3.68	28.3	2.45	18.8
2	Enforcement	2.26	17.4	2.29	17.6
3	Counseling and rehabilitation	1.32	10.2	1.83	14.0
4	Sanctions	1.08	8.3	1.76	13.6

dardized regression-based factor scores as these reflected the opinions of the relative efficacy of the countermeasure identified. Standardized regression-based factors, each with a mean of zero and a standard deviation of unity, would not make possible a meaningful comparison between the relative effectiveness of the four types of countermeasure. Cronbach's alpha coefficients for each of the composite scales were: Education = 0.76; Enforcement = 0.79; Counseling and Rehabilitation = 0.60; Sanctions = 0.62. These four scales were used in the following analysis.

Analysis of the Effectiveness of Drink-Flying Countermeasures

Analysis of the overall effectiveness of the four types of countermeasure using a repeated measures analysis of variance showed that there was a significant difference between theapproaches(F =121.65; df = 3,1329; p < 0.001).Tukey posthoc testsindicated that all means were significantly different from all others (= 0.01). Increasing the level of enforcement of the drink-flying regulations was regarded as the most effective countermeasure, followed (in order of perceived effectiveness) by levying sanctions on offenders, providing education about the effects of alcohol and the provision of counseling and rehabilitation programs (see Table 6).

To assess the potential effectiveness of the four categories of countermeasure with respect to offender and license category, a multivariate analysis of variance (MANOVA) was performed with effectiveness of countermeasures as the dependant variables (see Tables 6 and 7). Higher scores are indicative of greater perceived effectiveness in counteracting drinking and flying behavior.

TABLE 6
Mean, Standard Deviation, and Valid Number of Responses for Each of the Scale Scores
Derived From the Principle Components Analysis of the Education and Enforcement Survey
Items, Broken Down by Drinking and Flying Offender Category and License Type

License	Non-Drink Flyer			Non-Believer			Inadvertent Drink Flyer			Row Total		
	x	SD	n	x	SD	n	x	SD	n	x	SD	n
PPL												
Education	5.01	1.14	155	5.05	1.27	55	4.58	1.22	29	4.97	1.18	239
Counseling	3.78	1.39		3.80	1.45		3.20	1.00		3.71	1.37	
Enforcement	5.46	1.24		5.25	1.49		4.94	1.41		5.35	1.33	
Sanctions	5.51	1.25		5.17	1.41		4.82	1.30		5.34	1.31	
CPL												
Education	4.64	1.19	14	3.95	1.12	15	4.96	0.62	6	4.46	1.15	45
Counseling	3.96	1.27		3.98	1.02		3.83	1.11		3.95	1.15	
Enforcement	5.17	1.49		5.31	1.31		5.56	0.81		5.27	1.34	
Sanctions	4.93	1.58		5.02	0.89		3.50	1.22		4.77	1.42	
ATPL												
Education	4.64	1.35	90	4.12	1.46	17	4.64	0.81	21	4.42	1.37	188
Counseling	4.24	1.21		3.73	1.40		3.53	0.99		3.99	1.29	
Enforcement	5.31	1.22		4.64	1.62		4.95	1.47		4.97	1.45	
Sanctions	4.98	1.34		4.49	1.59		4.56	1.46		4.72	1.48	
Column total												
Education	4.85	1.23	269	4.46	1.43	147	4.64	1.02	56	4.70	1.28	422
Counseling	3.95	1.34		3.78	1.38		3.53	1.04		3.85	1.32	
Enforcement	5.38	1.36		4.93	1.56		4.95	1.38		5.19	1.39	
Sanctions	5.28	1.34		4.80	1.50		4.56	1.39		5.04	1.42	

Note. PPL = private pilot's licence; CPL = commercial pilot's licence; ATPL = airline transport pilot's licence.

TABLE 7

Multivariate Analysis of Variance Table: Countermeasures Factors Derived From Each of the Scale Scores From the PCA of the Education and Enforcement Survey Items, Analyzed by Drinking and Flying Offender Category Main Effect, License Type Main Effect, and Interaction Term

Drinking and Flying Offender Category	Λ	F	Effect df	Error df	p
Multivariate effcts					
Overall	0.06	2.43	8	864	<.05
Function 2 alone	0.99	1.91	3	433	ns

	DFA Coefficient (Function 1)	F	p		
Univariate effects (df = 2, 435)					
Countermeasure					
Education	0.37	2.53	<.10		
Counseling	-0.34	0.92	ns		
Enforcement	0.29	1.27	ns		
Scanctions	-1.00	5.81	<.01		

License Category	Λ	F	Effect df	Error df	p
Multivariate effects					
Overall	0.01	5.22	8	864	<.001
Function 2 alone	0.99	1.71	3	433	ns

	DFA Coefficient (Function 1)	F	p		
Univariate effects (df = 2, 435)					
Countermeasure					
Education	-0.66	3.85	<.050		
Counseling	0.81	2.70	<.100		
Enforcement	0.05	2.32	<.100		
Scanctions	-0.67	6.72	<.001		

Interaction Term	Λ	F	Effect df	Error df	p
Multivariate effects					
Overall	0.95	1.29	16	1,740	ns
Functions 2–4	0.98	1.05	9	1,054	ns
Functions 3–4	0.99	0.93	4	868	ns
Function 4 alone	1.00	0.06	1	435	ns

	F	p			
Univariate effects (df = 2, 435)					
Countermeasure					
Education	1.93	ns			
Counseling	1.23	ns			
Enforcement	1.25	ns			
Scanctions	1.05	ns			

Note.   PCA = principle components analysis; DFA = discriminant function analysis.

The results in Table 7 indicate there were differences in opinion about the effectiveness of the countermeasures with respect to offender category and type of license. No significant interaction term was observed. Standardized discriminant function coefficients are reported for significant functions.

With respect to offender category, decomposition of the multivariate term suggests that the principal difference between the groups lie in the respondent's opinions about the effectiveness of sanctions. The parameter estimates indicated that non-drink flyers regarded sanctions more likely to be effective than did the overall sample ($t = 2.99$; $p < 0.01$). Inadvertent drink flyers felt that sanctions would be less likely to be effective than the overall sample ($t = -3.13$; $p < 0.01$). There was also some suggestion that non-believers regarded sanctions unlikely to be an effective countermeasure ($t = -1.65$; $p < 0.10$). In addition, non-drink flyers tended to regard education more positively as a countermeasure than did the sample as a whole ($t = 1.61$; $p < 0.10$).

Analysis of the significant multivariate main effect with license type as the independent variable showed that three of its four constituent components were significant and one was approaching significance. The parameter estimates showed that private pilots felt that education and sanctions were significantly more likely to be effective countermeasures than did the overall sample (education, $t = 2.38$; $p < 0.05$ and sanctions, $t = 3.53$; $p < 0.001$). CPL holders tended to regard sanctions significantly less likely to be an effective countermeasure ($t = -2.00$; $p < 0.05$). There was some suggestion that holders of an ATPL were more likely to view counseling as an effective countermeasure than did the sample as a whole ($t = 1.45$; $p < 0.10$). ATPL qualified pilot's thought that both enforcement issues and sanctions were significantly less likely to be effective countermeasures (enforcement: $t = -2.09$; $p < 0.05$: sanctions $t = -3.13$; $p < 0.01$).

DISCUSSION

The results in Table 2 support the previous findings of Widders and Harris (1997) who observed that approximately 50% of U.K. pilots fell into either the non-believer or inadvertent drink-flyer offender categories and that holders of an ATPL were over represented in the non-believers category of potential offenders. Maxwell and Harris (1999), also found that possessing a professional pilot's license was a strong determinant of drink-flying offending. In this later survey, again over 50% of respondents reported that they had (or may have) flown when their BAC was in excess of the prescribed 0.02% limit giving further credibility to these results. These results stand in stark contrast to previous findings from a survey of U.S. pilot's that suggested that drinking and flying was thought more likely to be a problem with private pilots than professional pilots (Ross & Ross, 1992). An even earlier study found that only 3.5% of U.S. pilots admitted to flying after having consumed alcohol (Damkot & Osga, 1978).

The PCA performed on the opinions about the potential effectiveness of the various countermeasures produced four factors each relating to a different aspect of the tripartite approach described by Vingilis and Salutin (1980): at the primary level, the education of pilots concerning the effects of alcohol; at the secondary level, strategies for the enforcement of drinking and flying regulations; and at the tertiary level two factors were produced, one concerning the counseling and rehabilitation of pilots with personal problems and a second factor concerning the severity of sanctions for offenders (see Table 4).

An analysis of the scale scores derived from the PCA for the sample as a whole clearly indicated that U.K. pilots tended to view enforcement (secondary level) and sanctions (tertiary level) as the most effective drink-flying countermeasures (see Table 6). Education (the primary level intervention) was only rated as the third most effective remedial action. Counseling and rehabilitation was viewed as the least effective countermeasure. These results differ somewhat from those reported by Ross and Ross (1995) who found that in a sample of U.S. pilots, counseling was regarded as the most effective countermeasure, followed by education about the effects of alcohol. Ross and Ross (1995) found increasing the severity of sanctions was regarded as a relatively ineffective strategy. Studies of the general deterrence of drinking and driving, however, suggest that increasing the likelihood of apprehension when offending and implementing punitive sanctions can be effective countermeasures (Ross, 1988; Kinkade & Leone, 1992). It should also be noted, however, that the study by Ross and Ross (1995) could best be placed within a framework of *specific* deterrence rather than *general* deterrence, as many aspects of it examined the effectiveness of employee assistance programs in the elimination of drinking and flying behaviors.

There were differences in the opinions about the effectiveness of countermeasures between the 'non-drink flyers' and both the hypothetical offender groups (see Tables 6 and 7). Non-drink flyers felt that sanctions were more likely to be the most effective drink-flying countermeasure. Thus, sanctions were more likely to be regarded as an effective deterrent by the section of the flying population that was less likely to offend. The fact that sanctions were regarded as a less effective countermeasure by both offender groups compliments the findings from the drink-driving literature (e.g., Mäkinen, 1988; Ross, 1988; Wheeler & Hissong, 1988; McKnight & Voas, 1991). This observation should be kept in perspective, though, as the overall scale ratings for sanctions were quite high, second only to increasing the levels of enforcement of the drinking and flying regulations. It is interesting to note, though, that even inadvertent drink-flyers regarded the effectiveness of sanctions in a similar manner to non-believers.

Private pilots regarded education about the effects of alcohol as a potentially effective strategy to reduce the likelihood of offending behavior. The professional pilots in the sample were less convinced about the effectiveness of this strategy. It should be noted, though, that U.K.-licensed professional pilots are now required to pass a human performance and limitations examination, which incorporates the effects of alcohol on performance as part of the syllabus, hence, they may not perceive themselves as requiring more education in this area. Many airlines also supplement the guidance on alcohol consumption promulgated by the U.K. CAA with their own standing orders and advice for their aircrew. For example, several airlines require pilots to abstain from drinking alcohol for 24 hrs before reporting duty. Other operators require aircrew not to consume any alcohol at all for 8 hrs before reporting for duty or when on standby, and to consume not more than 5 units of alcohol (approximately 58g) in the 16 hrs preceding the 8 hr prohibited period.

The professional pilots in this study were less positively disposed toward tougher sanctions and enforcement compared to other respondents but tended to regard counseling and rehabilitation more favorably, thereby supporting the findings of Ross and Ross (1995), although it should be noted that these authors did not collect the opinions of private pilots. Even these sections of the sample, though, rated the effect of enforcement and sanctions more highly than the other countermeasures. It is possible that the more positive opinions of ATPL holders about the effectiveness of counseling and rehabilitation reflected the reasons why some respondents in this section of the sample consumed alcohol. Previous studies have found job-related pressures in professional aircrew to be a significant contributor to their drinking and flying behavior (e.g., Sloane & Cooper, 1984; Ross & Ross, 1995; Maxwell & Harris, 1999).

When comparing the results of this study and earlier surveys of U.K. pilots (Widders & Harris, 1997; Maxwell & Harris, 1999), to similar studies of U.S. pilots (e.g., Ross & Ross, 1992; Ross & Ross, 1995), there would seem to be major differences in the pattern of results. The U.K. studies suggest a higher rate of offending and different opinions concerning what would be the most effective countermeasures. It is not clear if these findings relate to cultural differences between the two populations or are responses to the different regulations. It can be speculated that the higher level of offending reported in the U.K. is a product of the much lower BAC prescribed in the revision to JAR OPS. Ross and Ross (1995) commented that in the U.S., the 0.04% rule has been criticized as being too lenient. It is interesting to speculate if the same pattern of results as those found in this study would be observed in U.S. aviators if the upper BAC limit was reduced to 0.02%, as in Europe.

SUMMARY AND CONCLUSIONS

As Ross and Ross (1995) noted, it is important to take account of pilots' opinions about what constitutes effective countermeasures as they have the most experience of the use of alcohol before flying and best know what would be most likely to deter them. In the U.K., the results from this survey indicate that efforts would be best placed in developing enforcement and sanctions countermeasures (the secondary and tertiary aspects of the tripartite model). These should primarily be aimed at holders of an ATPL, who are over represented in the non-believers offender group. Contrary to what may initially be expected (i.e., that educational countermeasures would be effective) these actions should also be of benefit in deterring offending behavior in inadvertent drink-flyers. It must be noted that the introduction of new drinking and flying regulations will not per se reduce the likelihood of offending. Five years after the 0.04% BAC rule was introduced by the FAA, only 37% of pilots were aware of its existence (Ross & Ross, 1990). To be effective, such regulations need to be complimented with propaganda if they are to have a deterrent effect. Cousins (1980) found that a propaganda campaign aimed at increasing the subjective probability of arrest had a beneficial effect when it was undertaken concurrently with the introduction of new sanctions for drink-drivers. The results from this study would suggest that antidrink flying propaganda placing emphasis on the sanctions that offenders could face and emphasizing high-profile measures that could be taken to enforce the regulations would be the most effective strategy to employ. In common with drinking and driving, random breath testing would seem to be the most effective single enforcement countermeasure that could be deployed.

REFERENCES

Albery, I. P., & Guppy, A. (1995) The interactionist nature of drinking and driving: A structural model. *Ergonomics, 38,* 1805–1818.

Balfour, A. C. J. (1988) Aviation pathology. In J. Ernsting and P. King (Eds.), *Aviation medicine* (2nd ed.). London: Butterworths, 703–709.

Berger, D. E., & Snortum, J. R. (1986). A structural model of drinking and driving: Alcohol consumption, social norms and moral commitments. *Criminology, 24,* 139–153.

Bierness, D. J. (1987). Self-estimates of blood alcohol concentration in a drinking-driving context. *Drug and Alcohol Dependence, 19,* 79–90.

Cousins, L. S. (1980). The effects of public education on the subjective probability of arrest for impaired driving: A field study. *Accident Analysis and Prevention, 12,* 131–141.

Damkot, D. K., & Osga, G. A. (1978). Survey of pilots' attitudes and opinions about drinking and flying. *Aviation, Space and Environmental Medicine, 49,* 390–394.

Department of Transport (1995). The Air Navigation (No. 2) Order. London: Her Majesties Stationary Office.

Drew, G. C., Colquhoun, W. P., & Long, H. A. (1959) *Effects of small doses of alcohol on a skill resembling driving.* Medical Research Council Memorandum No. 38, London: HMSO.

Dunbar, J. A., Penttila, A., & Pikkarainen, J. (1987). Drinking and driving: Success of random breath testing in Finland. *British Medical Journal, 295,* 101–103.

Flynn, C. F., Sturges, M. S., Swarsen, R. J., & Kohn, G. M. (1993). Alcoholism and treatment in airline aviators: One company's results. *Aviation, Space and Environmental Medicine, 59,* 314–318.

Guppy, A. (1988). Factors associated with drink-driving in a sample of English males. In, J. A. Rothengatter & R. A. de Bruin (Eds.) *Road user behavior: Theory and research.* Assen, The Netherlands: Van Gorcum, 375–380.

Hays, W. L. (1988). *Statistics* (4th ed.). New York: Holt, Reinhardt and Winston.

Homel, R., Carseldine, D., & Kearns, I. (1988). Drink-driving countermeasures in Australia. *Alcohol, Drugs and Driving, 4,* 113–144.

Kinkade, P. T., & Leone, L. M. C. (1992). The effects of 'tough' drunk-driving laws on policing: A case study. *Crime and Delinquency, 38,* 239–257.

Mäkinen, T. (1988). Enforcement studies in Finland. In, J.A. Rothengatter and R.A. de Bruin (Eds.) *Road user behavior: Theory and research.* Assen, The Netherlands: Van Gorcum, 584–588.

Maxwell, E., & Harris, D. (1999). Drinking and flying: A structural model. *Aviation, Space and Environmental Medicine 70,* 117–123.

McKnight, A. J., & Voas, R. B. (1991). The effect of license suspension upon DWI recidivism. *Alcohol, Drugs and Driving, 7,* 43–54.

Mertens, C. H., Ross, L. E., & Mundt, J. C. (1991) Young drivers' evaluation of driving impairment due to alcohol. *Accident Analysis and Prevention, 23,* 67–76.

Norström, T. (1978). Drunken driving: a tentative causal model. *Scandinavian Studies in Criminology, 6,* 252–283.

Norström, T. (1981). Drunken driving: a causal model. In, L. Goldberg (Ed). *Alcohol, drugs and traffic safety: proceedings of the 8th International Conference on Alcohol, Drugs and Traffic Safety,* Stockholm: Almqvist and Wiksel, 1215–29.

Peck, R. C. (1991). The general and specific deterrent effects of DUI sanctions: A review of California's experience. *Alcohol, drugs and driving, 7,* 13–42.

Ross, H. L. (1984). *Deterring the drinking driver: Legal policy and social control.* Lexington, MA: D.C. Heath & Co.

Ross, H. L. (1988). Deterrence-based policies in Britain, Canada and Australia. In, M. D. Lawrence J. R. Snortum & F.E. Zimring (Eds.) *Social control of the drinking driver.* Chicago: University of Chicago Press, 64–78.

Ross, L. R., & Ross, S. M. (1988). Pilot's attitudes towards alcohol and flying. *Aviation, Space and Environmental Medicine, 59,* 913–919.

Ross, L. R., & Ross, S. M. (1990). Pilots' knowledge of blood alcohol levels and the 0.04 percent blood alcohol concentration rule. *Aviation, Space and Environmental Medicine, 62,* 412–417.

Ross, L. R., & Ross, S. M. (1992) Professional pilots' evaluation of the extent causes and reduction of alcohol use in aviation. *Aviation, Space and Environmental Medicine, 63,* 805–808.

Ross, S. M., & Ross, L. R. (1995). Professional pilots. views of alcohol use in aviation and the effectiveness of employee assistance programs. *International Journal of Aviation Psychology, 5,* 199–213.

Russel, J. C., & Davis, A. W. (1995). *Alcohol rehabilitation of airline pilots* (FAA/DOT Report No. DOT/FAA–AM–85–12). Washington, D.C.: Federal Aviation Administration.

Sloane, H. R., & Cooper, C. L. (1984). Health-related lifestyle habits in commercial airline pilots. *British Journal of Aviation Medicine, 2,* 32–41.

Snortum, J. R., & Berger, D. E. (1989). Drink-driving compliance in the United States: perceptions and behavior in 1983 and 1986. *Journal of Studies on Alcohol, 50,* 306–319.

Vingilis, E. R., & Salutin, L. (1980). A prevention programme for drinking and driving. *Accident Analysis and Prevention, 12,* 267–274.

Wells-Parker, E., Bangert-Drowns, R., McMillen, R., & Williams, M. (1995). Final results from a meta-analysis of remedial actions with drink-driving offenders. *Addiction, 9,* 907–926.

Wheeler, G. R., & Hissong, R. V. (1988). Effects of criminal sanctions on drunk drivers: Beyond incarceration. *Crime and Delinquency, 34,* 29–42.

Widders, R. (1994). *Pilots' knowledge of the relationship between alcohol consumption and levels of blood alcohol concentration.* MSc Thesis, Cranfield University. Department of Applied Psychology.

Widders, R., & Harris, D. (1997). Pilots' knowledge of the relationship between alcohol consumption and levels of blood alcohol concentration. *Aviation, Space and Environmental Medicine, 68,* 531–537.

Manuscript first received August 2000

Appendix 2:
Harris, Chan-Pensley and McGarry (2005)

Harris, D., Chan-Pensley, J. & McGarry, S. (2005). The development of a multidimensional scale to evaluate motor vehicle dynamic qualities. *Ergonomics, 48*, 964–982.

Commentary

This paper was almost the converse in its conception to the one described in the previous appendix. In this case the work was an extension of earlier studies undertaken to develop a multidimensional aircraft handling qualities scale (Harris, Payne and Gautrey, 1999; Harris, 2000; Harris, Gautrey, Payne and Bailey, 2000). The objective was to describe the dynamic qualities of road cars (ride, handling, steering, etc) in such a way that they could be scaled meaningfully but it should also reflect that what is desirable in a luxury saloon car may not be so desirable in a sports car. This was done from the point of view of the end user (i.e. a normal driver) not a specialist test driver. Furthermore, there had been criticisms that many scales and techniques used in Ergonomics were neither psychometrically robust, reliable nor valid (see Stanton and Young, 1999a; b). For many subjective assessment techniques, including aircraft handling qualities scales, there was little (or no) reliability or validity data, thereby further supporting this observation.

The challenges in this paper were essentially three-fold. As before, the general ergonomics readership (or the readership from the car industry) had to be made aware of a literature that they would not normally peruse, that from the aviation domain (again – be aware of your readers). Secondly, there was not a great deal of previous literature to go on, which made describing the theoretical and practical context more difficult than usual. Thirdly, the final paper was a three-stage paper, put together from separate research studies, but which still had to read 'as a whole'. This required careful structuring to ensure that the reader was taken through the paper in such a way so they could appreciate how one study built upon the preceding study. In the end, it was structured as three short, almost self-contained studies (each comprising an Introduction; Method; Results; Discussion section) bracketed by a general Introduction and a general Discussion. Even though the paper is slightly longer than many papers (at around 8,500 words) there was still a lot to get in. As in Harris and Maxwell (2001) the treatment of the data was complex, requiring that the reader was led very carefully through the analyses. Lots of work was put in to provide a framework for what was to follow (see section 2 of the paper) combined with short summaries and further introductions (e.g. sub-sections 3.1 and 4.1) to act as *aide-memoirs*. Tell them what you are going to tell them; tell them it; tell them what you told them...

The 'Big Message'

This is a sensitive, multi-dimensional road vehicle dynamic qualities assessment technique, derived upon robust psychometric principles to ensure its validity.

The Story

- Vehicle dynamic qualities are multidimensional and have a strong interaction with environmental or task factors (in this case the type of road vehicle).
- Multidimensional scales have been developed in the aviation industry to describe aircraft handling qualities; the same approach can be used for cars.
- Many ergonomics scales have poor reliability and validity as a result of the way that they have been constructed.
- Study employs a multi-stage approach to ensure the validity and reliability of the dimensions to assess the vehicle dynamic qualities.

- The dimensions elicited (and the subsequent scale constructed) can discriminate meaningfully between different types of car in different categories.
- The results support the requirement for a multidimensional approach to the assessment of vehicle dynamic qualities and demonstrate the benefits of using a psychometrically robust approach to scaling.

Ergonomics, Vol. 48, No. 8, 22 June 2005, 964 – 982

The development of a multidimensional scale to evaluate motor vehicle dynamic qualities

DON HARRIS*, JAMIE CHAN-PENSLEY and SHONA MCGARRY

Human Factors Group, Cranfield University, Cranfield, Bedford, MK43 0AL, UK

Advances in motor vehicle engineering will allow greater refinement of the dynamic qualities of passenger cars in the near future. This paper describes the development and initial validation of a reliable and valid multidimensional scale to assess these parameters based upon a technique previously developed to evaluate aircraft handling qualities. The scaling methodology developed emphasizes the interaction between the vehicle's dynamic behaviour and the category of vehicle (e.g. sports car, executive saloon). This three-part study describes the initial extraction and the subsequent verification of the scale dimensions from an analysis of the opinions of *circa* 500 drivers, followed by an evaluation of the sensitivity and diagnosticity of the scale to distinguish between the road behaviour exhibited by vehicle types. The results suggest that the scale shows both content and construct validity, being able to distinguish both between broad categories of vehicle and dfferent models of vehicle within a particular category in a consistent and meaningful manner.

Keywords: Handling qualities; Driver behaviour; Subjective rating scales

1. Introduction

Modern technology has given the automotive engineer many more options from which to choose when optimizing the ride/handling qualities compromise of the modern motor vehicle. With the advent of increased computerization in road cars and suspension systems that are not dependent upon traditional steel springs (e.g. by using fast acting pneumatic or hydraulic systems) there will be even more ways available to refine their behaviour. To a limited extent, this is already happening in many 'top of the range' executive and sports cars, where drivers are given the option to select one of a number of pre-defined damper settings reflecting their preference for a sporting or comfort-oriented ride. More advanced systems becoming available also allow the driver an element of control over the vehicle's anti-roll bars, allowing them to decide upon the degree of roll exhibited by the car when cornering. However, with the introduction of computer-controlled hydro-pneumatic suspensions with no springs, the effective spring rates;

---

*Corresponding author. Email: d.harris@cranfield.ac.uk

Ergonomics
ISSN 0014-0139 print/ISSN 1366-5847 online © 2005 Taylor & Francis Group Ltd
http://www.tandf.co.uk/journals
DOI: 10.1080/00140130500181967

damping characteristics (both in bump and rebound); the roll characteristics of the vehicle (including roll centre); and anti-dive and anti-squat characteristics will all be able to be optimized through software changes. Furthermore, with the introduction of 'steerby-wire' systems (should changes in the design and build regulations allow them), other characteristics of the driver – vehicle interface will also be capable of being altered via simple changes to software settings, for example, steering gain and steering feel. The same will apply to other parameters, such as throttle response, which may be altered to make a car feel more sporting or more refined. In short, there is a potential revolution in the none-too-distant future concerning the manner in which the modern road car will ride and handle. Many of the current compromises imposed by current passively controlled mechanical systems will no longer apply.

With such flexibility becoming available concerning the dynamic behaviour of future vehicles, considerably more refined tools will be required to describe and evaluate the behaviour of these cars. Ideally, these descriptions should also be couched in the terms used by the end purchasers to describe their perceptions of their vehicle in everyday use. However, such a metric is not currently available. At present, the dynamic qualities of road vehicles tend to be evaluated by highly skilled test drivers using either very simple unidimensional handing qualities scales (e.g. that developed by the Society of Automotive Engineers 1985); by the qualitative comments of these experienced test drivers, typically evaluating the 'on-the-limit' aspects of the vehicle's behaviour; or by the measurement of engineering parameters related to parameters such as ride and handling. The unidimensional scales utilized have been pragmatically developed to fulfil in-company requirements rather than having been developed to produce psychometrically robust measures of ride and handling qualities. Qualitative comments from experienced test drivers are essential in refining a vehicle's behaviour. However, the inability to derive and quantify trends when using these scales as a result of their lack of diagnosticity, and inter-rater differences in opinion and assessment technique when using unidimensional scales ultimately limits their utility (see Field 1995, Harris 2000).

A directly analogous situation was faced by the aerospace industry several years ago with the advent of 'fly-by-wire' flight control systems. These fly-by-wire systems allowed the control engineer more and simpler ways by which to refine the handling qualities of aircraft compared to conventional technology mechanical flight control systems. The rating scale most commonly used at this time was the Cooper-Harper aircraft handling qualities rating scale (Cooper and Harper 1969). The Cooper-Harper scale was a simple, unidimensional scale ranging from 1 'Excellent, highly desirable' to 10 'major deficiencies: control will be lost. . .'. A fundamental problem with the use of this scale, however, was that the complexity of the aircraft's behaviour was not reflected in the dimensions of the scale (Payne and Harris 2000). In addition, the Cooper-Harper scale did not describe the interaction between the aircraft's dynamic qualities and the nature of the task being undertaken. What may be desirable qualities in one type of manoeuvre may not be so desirable in another type of manoeuvre (see Harris *et al.* 1999, 2000, Harris 2000). Similar criticisms may also be made of unidimensional scales to describe the handling qualities of motor vehicles. What may be a desirable characteristic in one type of vehicle (e.g. a more softly sprung, compliant ride in an executive car) may not be so desirable in another category of vehicle (e.g. a sports car). Unidimensional scales do not take into account the interaction between vehicle type, its dynamic road behaviour characteristics and its intended market segment.

It has also been noted in the aviation industry that when using simple physical measurements of an aircraft's behaviour to evaluate handling qualities a misleading impression may be gained (Harper and Cooper 1986). The engineer makes judgements about the aircraft's handling qualities from data derived from only a small part of the total pilot/vehicle system; individual parameters are analysed largely in isolation. Only the pilot can observe the complete, complex dynamic behaviour of the aircraft. Thus, these qualities cannot be defined simply by engineering descriptions of an aircraft's response to a pilot's control input. The authors concluded that it is essential that any description or assessment of handling qualities must include the pilot's perceptions of the behaviour of the aircraft. The same applies to road vehicles. The aim of vehicle performance, ride and handling evaluations is to assess if the dynamic qualities of the vehicle are suitable for the purpose and the market intended. To this end a vehicle is mostly evaluated by the subjective judgement of experienced researchers (Sano 1982).

As alluded to earlier, handling qualities are complex and multidimensional, yet this is not reflected in the scales to assess them. In the aviation domain, Payne and Harris (2000) identified five dimensions of aircraft handling that proved to be stable and could be easily related to the dynamic behaviour of the aircraft. These were subsequently developed into a multidimensional scale to assess aircraft handling qualities (the Cranfield Aircraft Handling Qualities Rating Scale (CAHQRS)). This scale, building upon concepts found in the National Aeronautics and Space Administration – Task Load Index (NASA – TLX) workload scale (Hart and Staveland 1988) also took into account the interaction between the aircraft's handling qualities and the task (Harris *et al.* 1999, 2000, Harris 2000). In unidimensional scales (such as the Cooper-Harper scale and the vehicle handling qualities scale proposed by Society of Automotive Engineers (SAE)) any interaction of handling qualities and the task and/or context are incorporated in the testing procedure rather than being a recorded, evaluated part of the measurement instrument.

Trials using CAHQRS showed it to have far greater test – re-test reliability than the Cooper-Harper scale. Overall, the test – re-test correlation for the full scale CAHQRS was 0.95; for individual sub-scales the test – re-test reliabilities ranged from 0.54 to 0.98 (Harris *et al.* 2000). In contrast the test – re-test reliability for the Cooper-Harper scale was in the region of 0.23 (Harris *et al.* 2000). This finding was in concordance with the earlier work of Wilson and Riley (1989) and Field (1995), who also observed that the Cooper-Harper scale exhibited poor inter-pilot reliability and that often the ratings provided did not tally with the opinions expressed on the pilots' comment cards. Furthermore, the CAHQRS also showed much greater diagnosticity. The greater reliability of the CAHQRS was attributed to the scale's format (and hence the method by which ratings were elicited). By using a simple, unidimensional format, first the Cooper-Harper scale implicitly requires the rater to form mentally a composite score based on all the individual aspects of the aircraft's handling. The rater is then required to weight these components, relative to one another, with regard to the requirements of the task being flown, before finally settling on a handling qualities rating. Many implicit judgements and comparisons are required prior to making the final rating, hence there is a great deal of scope for variability in several aspects of the process that can contribute to the lack of reliability. The CAHQRS makes this process explicit. Importance weightings for each of the components of aircraft handling for a particular manoeuvre are made explicitly prior to performing the task. The pilot is then required to make evaluations of five well-defined aspects of the aircraft's dynamic behaviour, irrespective of how important they are for the manoeuvre undertaken. Greater reliability is obtained as a result of making more than a single rating and also through the manner by which the various aspects of an aircraft's handling are combined to produce a rating, which is now through a mechanical and explicit process. In contrast, when using the Cooper-Harper scale the manner in which the various components of the way the aircraft's behaviour are combined is via subjective judgement and weighting to produce the final single score, which reflects a composite of several parameters.

Stanton and Young (1999a, b) have criticized the development of many ergonomics measures and methods for being neither psychometrically robust nor reliable. Indeed, for many often used scales in ergonomics, there is little (or no) reliability or validity data quoted (for example, see the compendium of human performance measures collated by Gawron 2000). Certainly all unidimensional measures of handling qualities can be criticized, as there is no evidence that they possess either construct or content validity. Using the guidelines from other areas in psychology, for example, personality theory or intelligence testing, the psychometric properties of many measurement scales used in ergonomics fall well below the acceptable norms for reliability, validity and measurement properties in general (for example, see Anastasi 1990 or Colle and Reid 1997).

For any scale of vehicle dynamic qualities to be useful (i.e. reliable, valid, diagnostic and sensitive to differences in appropriate vehicle parameters) there is a fundamental underlying assumption that drivers are sensitive to such aspects. There is, however, evidence that this is the case. Brindle (1984, 1986), using a subjective rating scale, demonstrated that drivers were sensitive to the differences between cross-ply and radial tyres. Furthermore, in a series of 'on the road' studies he also showed that drivers modified their driving behaviour in response to these tyre characteristics. More recently Walker *et al.* (2001) made similar observations showing that drivers were sensitive to the dynamic characteristics of different categories of vehicle.

This paper describes the development and initial validation of a multidimensional scale to assess road vehicle dynamic qualities. The method is based largely on that used to develop the CAHQRS. The objectives were to produce an instrument that described the manner in which everyday car drivers assessed the qualities of their vehicle (i.e. taking a user-centred/customer-focused approach). As Ike Iaconelli, Ford's director of Global Test Operations observed:

> One of the biggest problems that all manufacturers face is trying to replicate what happens in the real world, with real drivers. Our test drivers, for example, are skilled drivers who know how to get the best from a car and they cover more mileage than most. This, of course, does not make them representative of the typical Ford customer. . .'
>
> (Testing Technology International 1998).

The scale was required to have good content validity (i.e. its components must encompass all the pertinent aspects of the domain to be assessed); good construct reliability (i.e. the underlying constructs should be stable across samples); and it should have satisfactory indications of construct validity. Campbell (1960) suggested that an insight into construct validity could be gained via an assessment of a scale's convergent and discriminant validity. With regard to this latter aspect the scale was required to be able to distinguish between vehicle types in a reliable and consistent manner.

2. Overview of methodology

The method to develop the vehicle dynamics qualities rating scale progressed in three distinct phases. Phase One was concerned with eliciting the basic dimensions of vehicle ride, feel, performance and handling as described by a large sample of car drivers. Phase

Two essentially replicated Phase One in a further independent sample to ensure that the descriptions previously elicited were stable (i.e. the dimensions had construct validity). Phase Three transformed dimensions elicited into a rating scale to assess the dynamic qualities of road vehicles and evaluated the scale's sensitivity and discriminant validity (i.e. its ability to discriminate between categories and types of vehicle in a meaningful manner).

3. Phase One

3.1. *Initial elicitation of dimensions*

Approximately 400 adjectives to describe the dynamic qualities of passenger cars were collected from road tests published in the popular motoring press (e.g. Autocar). From these adjectives, the 52 most frequently used were identified and then paired with a word of diametrically opposite meaning to form a series of bi-polar adjective pairs. When constructing these bi-polar pairs it was always ensured that not only were the words antonyms of each other but also that there was one desirable and one undesirable pole. These bi-polar adjective pairs were then presented to 50 participants. These were recruited on an ad hoc basis but to be eligible they had to be in possession of a full UK driving licence. The participants were asked to evaluate the meaningfulness of these bipolar adjective pairs to describe the performance, feel, ride and handling qualities of a motor vehicle.

Initial qualitative analysis indicated that respondents had some diffculty in separating haptic feedback (such as steering 'feel') from the actual performance, handling and ride qualities of their vehicle. As it was desired to produce a customer-centred description of a vehicle's dynamic qualities and the aim was to produce a scale with high content validity it was decided to retain these items. For any scale with psychometric properties to have content validity, the items in it must evaluate all major aspects of the domain to be assessed, which requires a systematic assessment of that domain at the outset of the scale's development (Anastasi 1990).

After the initial parsing of the adjectives to identify the most applicable descriptions of vehicle road behaviour, the remaining 33 pairs were made into a short self-completion questionnaire. The order that the adjectival pairs were presented on the instrument was randomized. Each item used a 1 – 5, tick-box rating scale to indicate, in the opinion of the person completing the questionnaire, which adjective pole they agreed with most and to what degree. In approximately half of the cases the desirable pole was presented first and vice versa. Brief demographic details of the respondent (age, sex and number of years that they had held a full driving licence) and details of their motor vehicle(s) were also collected.

These questionnaires were distributed to UK motorists personally by one of the researchers in the car parks at a large UK sporting event. They were given a pack containing a covering letter about the aims and objectives of the study; instructions for completing the questionnaire and a FREEPOST envelope for its return. The respondents were asked to assess the ride and handling qualities of the vehicle that they used most frequently. In total 300 questionnaires were distributed.

3.2. *Analysis of responses*

Two hundred and twenty-three completed survey instruments were returned, representing a 78% response rate. Of these 181 (81.2%) were from male drivers and 42 (18.8%) from females. The age range of respondents was from 17 to 65 years (with a mean of 35.8

(SD 10.8) years) and the mean number of years of driving experience was 17.3 (SD 10.5). Five distinct categories of vehicle were represented in the sample as a whole, including 25.0% small cars (5 4.00 m); 44.4% medium cars (4.01 – 5.00 m); 22.2% large, executive cars (4 5.01 m); 2.8% off-road (four-wheel drive) vehicles, and the remaining 5.6% were classified as either sporting coupes or roadsters.

Prior to analysis, the data were transformed (as appropriate) so that in all cases a high score represented a desirable dynamic quality of the vehicle. The data were then subject to principal components analysis (PCA) using SPSS (version 9.00). From the PCA seven components were extracted accounting for 65% of the total variance in the sample. Kaiser's criterion (components with an Eigenvalue in excess of unity) was used to determine the maximum number of components extracted. These components were subject to a Varimax rotation to produce a 'cleaner', more interpretable solution. For an item to be deemed to load onto one of the extracted components it was required to have a loading in excess of 0.45 (significant p 5 0.01 – see Hair *et al.* 1998). Items loading onto two (or more) components were excluded from the final solution.

The final, post rotation principal components solution is described in table 1 along with its associated summary statistics in table 2. As can be seen, the principal components extracted largely distinct and recognizable descriptions of the ride, handling and steering qualities of motor vehicles.

The components extracted were then subject to an examination of their internal consistency (reliability). Component 1 ('steering qualities') had a Cronbach's alpha of 0.91. Component 2 ('performance') initially also had an alpha value of 0.91; however, two items ('rapid/ slow' and 'exciting/dull') actually were detrimental to the overall internal consistency of this component. Once these were removed, the alpha value was elevated to 0.95. Removal of these items also resulted in a more coherent and parsimonious dimension, in terms of its constituent components. Analysis of the internal consistency of 'ride composure' (Component 3) resulted in a Cronbach's alpha of 0.80 and Component 4 ('handling qualities') had a corresponding value of 0.78. The final two dimensions (Component 5, 'ride comfort' and Component 6 'grip') had Cronbach's alpha values of 0.77 and 0.83, respectively. All the values for Cronbach's produced were well over the minimum acceptable level of 0.7 for scale internal reliability as suggested by Cronbach (1951) and Robinson *et al.* (1991). As the final component extracted in the initial analysis was composed of only a single item, this dimension was not used in the following verification stage, as single item components are inherently unreliable.

4. Phase Two

4.1. *Verification of dimensions*

The objective of the second phase was to cross-validate the latent structure underlying vehicle dynamic qualities elicited in the PCA in Phase One in an independent sample to assess its stability, reliability (internal consistency) and construct validity. Only if these factors were satisfactory would the items elicited in Phase One be suitable for forming the basis of a psychometrically robust, multidimensional vehicle dynamic qualities rating scale.

All the items describing each of the six dimensions that remained after the analysis of the components internal consistency (see table 1) were formed into a second questionnaire using an identical response format to that described previously. The same demographic and vehicle data as before were also collected.

Table 1. Post Varimax rotation principal components solution (using Kaiser's criterion for component extraction) to describe the underlying dimensions of vehicle dynamic qualities

Bipolar item*	Principal component loading						
	1	2	3	4	5	6	7
Communicative/Vague	0.79						
Informative/Uninformative	0.75						
Accurate/Inaccurate	0.73						
Precise/Imprecise	0.71						
Responsive/Unresponsive	0.71						
Sensitive/Insensitive	0.7						
Interactive/Uninvolving	0.68						
Well-weighted/Poorly weighted	0.50						
Quick/Slow		0.89					
Speedy/Leisurely		0.87					
Frisky/Sluggish		0.84					
Good acceleration/Poor acceleration		0.80					
Rapid/Slow		0.67					
Exciting/Dull		0.66					
Predictable/Unpredictable			0.67				
Controlled/Uncontrolled			0.60				
Stable/Unstable			0.53				
Solid/Loose			0.53				
Composed/Fidgety			0.50				
Firm/Soft			0.45				
Settled/Unsettled				0.68			
Firm/Bouncy				0.58			
Poised/Nervy				0.57			
Oversteer/No oversteer				0.54			
Body roll/No body roll				0.53			
Taut/Slack				0.52			
Understeer/No Understeer				0.50			
Absorbent/Thumpy					0.78		
Smooth/Harsh					0.77		
Comfortable/Uncomfortable					0.69		
Grippy/Skiddy						0.68	
Adhesive/Slippy						0.66	
Light/Heavy							0.89

*Items are presented in order of magnitude of loading with the principal component and, for clarity, items with a loading below 0.45 have been omitted from the table.

Table 2. Principal component extraction summary statistics for analysis extracting the underlying dimensions of vehicle dynamic qualities

Component number	Component name	Eigenvalue	% of variance pre-rotation	% of variance post-rotation
1	Steering qualities	11.83	34.8	15.7
2	Performance	2.87	8.4	14.3
3	Ride composure	2.01	5.9	10.5
4	Handling qualities	1.82	5.4	8.8
5	Ride comfort	1.28	3.8	8.0
6	Grip	1.20	3.5	4.7
7	Unnamed	1.06	3.1	3.7

This second, self-completion questionnaire was distributed to 1000 drivers; however, in this case the categories of vehicle identified in the previous results section were specifically targeted to ensure that the cross validation sample was of a similar composition to the previous sample. Questionnaires were distributed to drivers in public car parks to aid in this respect.

As before, completed survey instruments were returned to the researchers using the FREEPOST envelope included in the questionnaire distribution pack.

4.2. Analysis of responses

Of the 1000 questionnaires distributed, 224 (22.4%) were returned in time for analysis, comprising146 (65.2%) from male drivers and 78 (34.8%) from female drivers. The age range of respondents was from 18 to 68 years (with a mean of 41.7 years). The final sample of vehicles was composed of 22.3% small cars, 37.5% medium cars, 27.7% large cars and 12.5% sporting coupes and roadsters.

Prior to analysis, the data were treated in exactly the same manner as that described at the beginning of the previous results section. The data were then subject to a confirmatory factor analysis (CFA) using AMOS 4.1. A maximum likelihood method was selected for factor extraction. The hypothesized factor structure, against which the data in Phase Two, were compared was that derived in the PCA in Phase One.

An initial CFA was only a moderately good fit to the hypothesized structure of road vehicle dynamic qualities elicited in Phase One (w^2 minimum = 753.68; df = 362; p 5 0.001: adjusted w^2 = 2.08: 'goodness of fit index' = 0.81: 'adjusted goodness of fit index = 0.77). However, the deletion of several items (listed in table 3) on the basis of the computed modification indices (the likely reduction in the overall w^2 goodness-of-fit value) produced a slightly modified model of indicator variables loading onto latent factors, which produced a model with an extremely good fit to the hypothesized latent structure of vehicle behaviour (w^2 minimum = 150.97; df = 120; p 5 0.03: adjusted w^2 = 1.26: 'goodness of fit index' = 0.93: 'adjusted goodness of fit index = 0.90). All these values are well in excess of those suggested by Hair *et al.* (1998) as being indicative of an acceptable CFA solution. It is likely that in the initial extraction of components (described in Phase One and in tables 1 and 2), certain variables loading onto a component were included as a result of capitalizing on chance associations within the

Table 3. Items deleted from the final confirmatory factor analysis solution and their associated principal components

Bipolar item	Component name
Communicative/Vague	Steering qualities
Informative/Uninformative	Steering qualities
Precise/Imprecise	Steering qualities
Sensitive/Insensitive	Steering qualities
Well-weighted/Poorly-weighted	Steering qualities
Good accleration/Poor acceleration	Performance
Predictable/Unpredictable	Ride composure
Composed/Fidgety	Ride composure
Settled/Unsettled	Handling qualities
Taut/Slack	Handling qualities
Comfortable/Uncomfortable	Ride comfort

data set rather than as a result of reflecting the true underlying dimensions of vehicle dynamic qualities. As a result, these variables are highly likely to be eliminated in the cross-validation phase. It is also noticeable that the majority of variables deleted were from the larger components in the initial PCA. This may suggest that these components were what Cronbach (1951) has termed a 'bloated specific' (i.e. a highly specific description of a very narrow underlying construct). This was not reflected in the second administration of the scale, which used a slightly smaller sub-set of items (see discussion in section 3.2.). The final model also served to increase the parsimony of the solution and enhance the interpretability of the factors produced. The final factor structure from the CFA is described in figure 1 and the inter-correlations between the latent variables are contained in table 4.

As it was proposed to use the items from each of the latent variables in the CFA to form the basis of a scale for constructing a multidimensional vehicle dynamic qualities rating scale, the internal consistency of these factors was again assessed to ensure that the items were all good indicators of the same underlying construct. The results of this

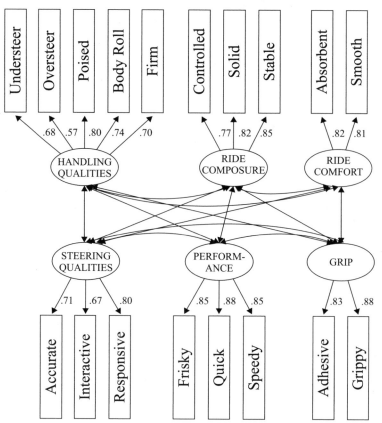

Figure 1. Final Confirmatory factor analysis factor structure to describe vehicle dynamic qualities. The numbers above the straight arrows represent the standardized regression weights between the latent variable and the indicator variables. Values for the inter-correlations between the latent variables can be found in table 4.

Table 4. Values for the inter-correlations between the latent variables (Pearson's r) describing vehicle dynamic qualities

Component name	Steering	Performance	Grip	Comfort	Composure
Performance	0.57				
Grip	0.84	0.49			
Ride comfort	0.42	0.20	0.39		
Ride composure	0.77	0.53	0.69	0.45	
Handling qualities	0.84	0.56	0.79	0.44	0.83

analysis using Cronbachs' alpha can be found in table 5. Once again it can be seen that all the values for Cronbach's alpha were well over the minimum acceptable level suggested by Cronbach (1951) and Robinson *et al.* (1991) of 0.70 for internal consistency.

4.3. Discussion of results

The results from the CFA performed on the data from the independent, cross-validation sample suggest that the dimensions of vehicle dynamic qualities elicited are stable, replicable and thus have construct validity. The manner by which the items were originally selected for inclusion in the study in Phase One should also ensure their content validity. The Cronbachs' alpha statistics computed for each dimension would also suggest that the dimensions are internally consistent (reliable).

Although the cross-validation of dimensions in the CFA has resulted in some high inter-correlations amongst the factors, this is to be expected as structural equation modelling makes no attempt to produce orthogonal factors (cf. the Varimax rotation used in Phase One). Inspection of the inter-correlations between the dimensions, however, does suggest that drivers can distinguish between various aspects of their vehicles' dynamic behaviour and make separate evaluations of them. For example, the majority of the correlations involving the dimensions of 'ride comfort' and 'performance' are all relatively small (see table 4). However, the correlations between 'steering qualities' and both 'grip and handling qualities' are high, suggesting that drivers regard these dimensions as being highly inter-related in everyday driving.

On the basis of these results a multidimensional road vehicle dynamic qualities rating scale was constructed using similar concepts to those found in the NASA – TLX mental workload scale (Hart and Staveland 1988) and the CAHQRS multidimensional aircraft handling qualities rating scale (Harris *et al.* 2000).

5. Phase Three

5.1. Scale construction and data collection

There were two basic components to the construction of the rating scale, the scale evaluating each component of vehicle dynamic behaviour for its adequacy and a further component that rated the importance of each of these aspects with regard to the role/ category of the vehicle (q.v. the weighting index of the NASA – TLX or the criticality index of the CAHQRS).

The rating aspect of the scale was constructed simply from a mean of the ratings on all the components making up each sub-scale (as derived from the final CFA) once the items had been transformed to ensure that a high rating represented a desirable ride

Table 5. Cronbach's alpha values for the internal consistency (reliability) of the latent
variables describing vehicle dynamic qualities

Component name	number of items	Cronbach's alpha
Performance	3	0.90
Steering qualities	3	0.78
Grip	2	0.86
Ride comfort	2	0.80
Ride composure	3	0.86
Handling qualities	5	0.83

or handling quality (exactly as the data were treated in Phases One and Two). The ratings aspect of the scale was presented in the exactly same manner as previously described.

After completing the rating scales, respondents were also required to rank the importance of each dimension of the vehicle's dynamic qualities with regard to the category of the vehicle being rated. The importance scale comprised a ranking from 1 (most important) to 6 (least important). Short definitions of the vehicle dynamic qualities dimensions can be found in table 6.

To test the sensitivity and predictive validity of the scale it was administered to drivers of: 49 'superminis' (e.g. Ford Fiesta and Peugeot 206); 48 small hatchbacks (e.g. Ford Focus and Volkswagen Golf); 33 medium-sized cars (e.g. Ford Mondeo and Vauxhall/ Opel Vectra); 61 small executive cars (e.g. Audi A4 and Mercedes-Benz C-Class) and 22 small, sporting coupes (e.g. Ford Puma and Renault Meganee Coupe). To target the drivers of the specific vehicle types required for the study, drivers were approached directly in the car parks of a major UK shopping centre as they were returning to their vehicles. Within each of these sub-samples it was ensured that there was a reasonably large sample of at least two models, including the class-leading vehicle (as defined in Autocar magazine) at the time of the study.

5.2. *Treatment of data and scale scoring*

For the sake of consistency and to aid interpretation, the scale score data were reversed so that a low score represented a desirable dynamic characteristic (i.e. scores were presented in the same fashion as the Cooper-Harper scale and other scales using this format, such as the Bedford Scale by Roscoe (1984) and the Haworth-Newman Display Readability Scale, see Newman and Greeley 2001). The ranking data reflecting the importance of each of the dimensions for a particular category of vehicle were left unchanged.

Sub-scale scores reflecting a vehicle's dynamic qualities were calculated from the product of the mean scale value. For use in a diagnostic context, these scale scores were then multiplied by their associated importance ranking. Both the NASA – TLX workload scale and the CAHQRS use this approach. By taking this approach the sensitivity of the scale is enhanced (compared to a unidimensional scale) by 'gearing' the scale ratings by their relative importance to the category of vehicle. When making comparisons of vehicles within a category, however, this is not necessary, as described in the following section.

5.3. *Results of validation exercise*

5.3.1. Ranking of importance of the dimensions of vehicle dynamic qualities by category of vehicle. One of the basic premises of the scale under development is that

Table 6. Short definitions of the vehicle dynamic qualities dimensions used in the motor vehicle dynamic qualities rating scale

Scale component	Short definition
Performance	This involves the power of the vehicle and is typically reflected by its ability to accelerate
Steering qualities	This refers to the feedback supplied via the steering wheel. A good steering system should give a crisp and accurate response at the start of a corner and respond proportionally afterwards
Grip	This refers to the absolute lateral grip of the vehicle as a result of the adhesion of the tyres to the road surface
Ride comfort	This refers to the evaluation of the level of comfort when travelling over various road surfaces
Ride composure	This refers to the manner in which the body of the vehicle settles and rides over the road surface
Handling qualities	This refers to the manner by which the vehicle responds to the inputs from the driver

different aspects of a vehicle's dynamic qualities will be of differing levels of importance with regard to the vehicles category (e.g. sports car or executive car).

The rankings of the importance of each aspect of a vehicle's dynamic behaviour suggest that the relative importance of each aspect is different with regard to the category of the vehicle. For all categories of vehicle, with the exception of medium-sized saloon cars, a within vehicle category analysis of the ranks awarded (Friedman's analysis of ranks) shows significant differences with regard to the category of dynamic behaviour assessed (supermini: $\chi^2 = 24.64$, df = 5, $p < 0.001$; small hatchback: $\chi^2 = 46.87$, df = 5, $p < 0.001$; medium-sized car: $\chi^2 = 1.57$, df = 5, $p > 0.05$; small executive car: $\chi^2 = 22.97$, df = 5, $p < 0.001$; and small coupes: $\chi^2 = 13.53$, df = 5, $p < 0.02$).

A brief analysis of the rankings in table 7 suggests that (not surprisingly) drivers of small, sporting coupes regarded handling, steering and grip as being more important than ride quality. Conversely, drivers of executive cars emphasized ride composure and comfort over performance and handling qualities. The pattern of results from drivers of the smaller two categories of vehicle showed a far less clear pattern of rankings, although in both cases performance was placed at a premium. The pattern of rankings for medium-sized vehicles should be interpreted with caution, as there was very little difference in the mean ranks of the categories, suggesting no clear preference (overall) in the dynamic qualities of these vehicles. This may reflect the multi-purpose character of these cars.

5.3.2. Sub-scale validity and sensitivity. For the individual sub-scales to be valid they must be able to discriminate meaningfully between vehicles. To be useful, the scales must also be sensitive. Scales that discriminate only between the dynamic qualities of very disparate categories of vehicle (for example, off-road vehicles and sports cars) are of very limited utility.

To test the sensitivity of the sub-scales, the ratings of dynamic behaviour of one or more models within each category was compared to the class-leading vehicle. For the sake of brevity only two such analyses are reported and discussed here. As the objective in this case was to test the sensitivity of the sub-scales within a group, sub-scale scores were not multiplied by the corresponding importance ranking. These latter aspects reflect the importance of that component of dynamic behaviour within a class of vehicle, not between types of vehicle.

Table 7. Within vehicle category ranks for the importance of each aspect of vehicle dynamic qualities (1 = most important; 6 = least important)

Class	n	Performance	Steering qualities	Grip	Ride comfort	Ride composure	Handling qualities
Supermini	50	2	3	5	1	6	4
Small hatchback	47	1	6	3	5	4	2
Medium saloon	36	5	3	1	4	2	6
Small executive	62	5	3	6	2	1	4
Samll coupe	23	4	2	1	6	5	3

While it is moderately straightforward to demonstrate the sensitivity of the scales in this manner, it is far more difficult to demonstrate their criterion validity, as there is no readily measurable parameter to compare the scale results against. As a result, in an attempt to demonstrate criterion validity, the results were compared to comments taken from widely published vehicle road tests.

In a comparison of small hatchbacks, the class-leading vehicle (car A) was compared against the best selling similar vehicle in Europe (car B). These results are shown in table 8.

The results in table 8 show that the class leading vehicle was superior to its comparison vehicle in all respects, significantly so in terms of the performance and handling qualities dimensions of dynamic behaviour. The results for the steering qualities and ride composure dimensions are also verging on significance. This would suggest that these sub-scales are sensitive enough to differentiate between these two vehicles. However, the scales must differentiate between the cars in a meaningful way if they are to be valid. Of car A, What Car? magazine stated:

> . . .it's still hard to take in just how far ahead of the game What Car?'s small hatchback of the year has taken the driving pleasure you can expect from this class of car.
>
> You notice straight away how stable and composed [car A] feels and how every steering wheel movement produces an accurate response from the front wheels.
>
> Cornering is crisp and accurate. . . it rides badly pock-marked B-roads firmly and it's never harsh and refinement is terrific.
>
> (What Car? 2001)

Whilst it should be noted that these comments are taken from a motoring magazine, they do suggest that the ratings on the sub-scale dimensions of steering qualities, ride composure and handling qualities appropriately reflect the dynamic behaviour of the vehicle in question. There is further evidence in this respect when the corresponding reviews of car B are considered.

> The [car B] has had some of its thunder stolen by [car A] in the past year
>
> We have no complaints about the [car B's] overall competence but it won't delight a driver in the same way as the [car A] which stamps its authority over [car B] on the road

Table 8. Comparison of two leading small hatchbacks in terms of their rated dynamic qualities*

	Car	n	Mean	SD	t	p
Performance	A	24	1.92	0.53	-2.27	0.03
	B	24	2.46	1.04		
Steering qualities	A	24	1.71	0.49	-1.74	0.08
	B	24	2.04	0.80		
Grip	A	24	1.75	0.63	-0.88	0.38
	B	24	1.92	0.69		
Ride comfort	A	24	2.02	0.67	-1.27	0.21
	B	24	2.31	0.91		
Ride composure	A	24	1.64	0.43	-1.84	0.07
	B	24	1.99	0.82		
Handling qualities	A	24	1.97	0.50	-2.50	0.02
	B	24	2.48	0.86		

*Low figures, on a range of 1 – 5, indicate superior dynamic behaviour. t = t-value; p = associated probability.

> The [car B] has a little too much body roll and a lack of steering feel through bends . . . It can't boast a composed ride either.
>
> . . .copes with undulations and crests well but feels fidgety over potholes.
>
> (What Car? 2001)

These further comments would indicate that in terms of the relative sub-scale scores awarded by respondents, the scales are sensitive enough to differentiate between the two vehicles in terms of their dynamic responses and also do so in a meaningful, valid manner.

In a comparison of medium-sized cars, the class-leading vehicle (car C) was compared against its leading competitor in the UK fleet and medium-sized car market (car D). These results can be seen in table 9.

Table 9 shows that car C was rated as being superior in the areas of performance, steering qualities grip and there was also a strong suggestion that it was also better than car D in terms of its ride composure. Road test comments about car C support the results in table 9.

> Wonderfully crisp steering feel with a strong, self-centring action.
>
> (Autocar 1996)

> The [car C] has been a keen handler from day one and the facelift a couple of years ago only served to sharpen its responses. This ability to supply driver enjoyment on demand may be at the expense of ultimate ride comfort, but you have to drive over some very rutted road to pick up on this.
>
> (What Car? 2001)

Reviews of car D supported the results of the sub-scale comparisons with car C, providing further for the validity of the sub-scales. For example:

> Comfort over distance is the [Car D's] strength, not B-road entertainment. However, it's relatively pleasing to punt along, thanks to smooth engines and a

Table 9. Comparison of two leading medium sized cars in terms of their rated dynamic qualities*

	Car	n	Mean	SD	t	p
Performance	C	11	1.82	0.23	-3.69	0.00
	D	10	2.60	0.66		
Steering qualities	C	11	1.67	0.37	-2.01	0.06
	D	11	2.36	1.09		
Grip	C	11	1.66	0.45	-2.07	0.05
	D	11	2.45	1.23		
Ride comfort	C	11	1.82	0.56	-1.25	0.23
	D	11	2.23	0.93		
Ride composure	C	11	1.67	0.47	-1.77	0.09
	D	11	2.36	1.22		
Handling qualities	C	11	1.82	0.45	-1.54	0.16
	D	10	2.30	0.99		

*Low figures, on a range of 1 – 5, indicate superior dynamic behaviour.

> comfortable ride. But the [car D] is not as entertaining as a [car C or list of other competitors].
>
> (Autotrader 2002)

> Happiest and most refined on motorway; less so on twisty roads where handling and ride lack poise and polish.
>
> (Parker's 2002)

> Not a car to drive for the pleasure of it. . . (Parker's 2002)

One of the more interesting points to note is that the ratings were collected from member's of the general public about their current cars. In many cases (although it is impossible to determine how many) this is likely to be a judgement of the dynamic qualities of their vehicle in isolation, whereas many of the comments made by the motoring press are comparative comments from road testers who have had the benefit of driving all the cars compared in the previous two analyses. However, despite the respondents probably not having the benefit of comparison, the scales still discriminate between the models in the manner expected, again suggesting further evidence of validity.

5.3.3. Multidimensional presentation of scale results. In the present application, low scores and low rankings are more desirable or more important. As was also the case with the NASA – TLX and the CAHQRS, it was also possible to compute an overall scale score for the ride/handling qualities of the vehicle through the weighted summation of the individual sub-scale scores, although it is argued that this is of limited utility as the power of the scale resides in the interpretation of its component sub-scales.

Perhaps the greatest power of multidimensional scales lies in the graphical presentation of scores. When presenting data in this manner, in accordance with the practice employed by both the NASA – TLX and the CAHQRS, scale ratings of vehicle dynamic qualities were plotted (individually) on the y-axis and the importance rankings on the x-axis. Short histogram bars are desirable, and narrow bars indicate higher levels of importance. This approach was chosen as it emphasizes the interaction between the compromises made between the various dynamic qualities and the category of the vehicle. An example of this method of depicting the results is given in figure 2, where a poorly rated sports coupe (car X) is contrasted with a highly rated executive saloon (car Y). It can be seen from the width of the bars in the graphical representation that the dimensions of ride comfort and ride composure are regarded as being the most important for drivers of executive saloons, whereas drivers of sporting coupes place greater emphasis on grip, steering and handling qualities. The drivers of the class-leading car Y rated all aspects of their vehicle's dynamic properties superior to those of the old, poorly regarded small, sporting coupe, as indicated by the shorter histogram bars. Comments from the road tests concerning the dynamic behaviour of car X include:

> . . .it lacks the sophistication to really be pushed on. The steering is a bit too light at speed, and the roadholding's not secure enough under pressure. It's perfectly adequate for most, undemanding, drivers, but serious sports fans will be frustrated and disappointed
>
> (Autotrader 2002)

Besides being cramped, the [car X] also serves a bumpy, noisy ride.

(Autotrader 2002)

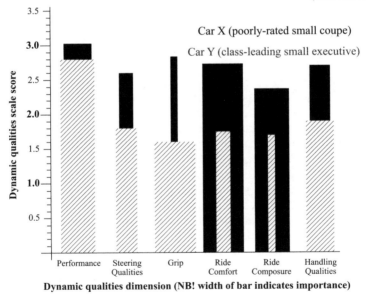

Figure 2. Example of the graphical presentation of the multidimensional vehicle dynamic qualities rating scale contrasting a poorly rated sports coupe with a highly rated executive saloon. NB: Lower ratings indicate more desirable dynamic behaviours and narrow histogram bars indicate higher levels of importance of a particular dimension with regard to that vehicle's market sector. To interpret a dynamic qualities profile it should be remembered that if a bar is narrow it should also be short.

These comments would seem to endorse the ratings awarded by its drivers. In contrast, the comments for car Y tend to be more complimentary, y*et al*so reflect its different role and the ratings awarded by its drivers:

> . . . good ride and remarkable smoothness at cruising speeds.
>
> (What Car? 2001)

> High marks here – the [car Y] is a very relaxed and secure motorway cruiser, and has better ride quality than most in the class.
>
> (Autotrader 2002)

> Performance – Not a strong point. . .
>
> (Autotrader 2002)

6. Discussion

The scale developed to assess motor vehicle dynamic qualities was based upon scaling principles initially used in the NASA – TLX multidimensional workload scale (Hart and Staveland 1988) and the Cooper-Harper aircraft handling qualities scale (Cooper and Harper 1969), which where subsequently developed in the CAHQRS (see Harris *et al.* 2000). Using the same approach as in the later case, a scale has been developed that allows everyday drivers to assess meaningfully the dynamic behaviour of their vehicles. The aims were to produce a scale that had good content validity; good construct reliability and which was sensitive to differences between the dynamic behaviour of different vehicles. Through the manner in which the scale was initially developed the content validity should be assured. The replicability of the underlying factor structure (figure 1) is also indicative of good construct validity (see Anastasi 1990).

The establishment of criterion reliability for abstract concepts such as the assessment of the dynamic qualities of a motor vehicle is much harder to establish (q.v. the handling qualities of an aircraft – see Harris *et al.* 2000). The broad agreement of the scale ratings with the opinions of professional motor-magazine road testers does suggest that the scale aids in producing valid ratings of the dynamic behaviour of motor vehicles (see tables 8 and 9 and figure 2).

The use of a NASA – TLX type of scale format once again shows the importance of being able to rate and display the interactive nature of scale ratings with context. Previously, in the case of the CAHQRS, this interaction was between aircraft handling qualities and the nature of the manoeuvre being flown. In the present case the interaction is between the category of motor vehicle and its dynamic behaviour. The differences in the rankings of importance of the various aspects of a vehicle's dynamic behaviour indicate that the drivers of the five different categories of vehicle included in this study did not all want the same things of their vehicles (see table 7). This underlines the importance of taking the assessment approach suggested herein. Simple ratings of dynamic qualities on each of the sub-scales are not enough. To draw conclusions simply from these assessments would overlook the complex interaction with the category of the vehicle and hence the relative importance of each aspect of its behaviour.

It is again interesting to note that the everyday driver is sensitive to differences in the dynamic behaviour of road vehicles, which supports the earlier observations of Brindle (1984, 1986) and Walker *et al.* (2001). It is not just test drivers that are sensitive to these aspects, which further strengthens the call to involve everyday drivers in the development of motor vehicles' on-the-road dynamic qualities. There is much more scope for a user-centred approach in this area than may have been initially thought.

Further development of the scale is planned in a series of trials in an engineering simulator and in a set of test-track trials. These trials will further evaluate the criterion validity of the scale and also assess its test – re-test reliability and sensitivity. However, the initial results would seem to suggest that there is a sound basis on which to continue further scale development and assessment, thereby avoiding the criticism of other measures in ergonomics that they do not conform to even the most basic standards for psychometric measurement.

References

ANASTASI, A., 1990, *Psychological Testing*, 6th edition (New York: MacMillan).

AUTOCAR, 1996, *Road Test* 4129. 4 December, 58 – 59.

AUTOTRADER, 2002, Online road test reviews. Available online at: http://ces.autotrader.co.uk/ces/search. jsp (accessed 11 March 2005). BRINDLE, L.R., 1984, The influence of tyre characteristics on driver opinion and risk-taking, PhD thesis. College of Aeronautics, Cranfield University, Cranfield, UK.

BRINDLE, L.R., 1986, Aspects of subjective/objective correlation regarding vehicle tyres. In *Proceedings of the Tenth International Technical Conference on Experimental Safety Vehicles*, Oxford, England, 1 – 4 July 1985 (Washington, DC: US Department of Transportation, National Highway Traffic Safety Administration). Report number DOT HS 806 916, February 1986.

CAMPBELL, D.T., 1960, Recommendations for APA test standards regarding construct, trait and discriminant validity. *American Psychologist*, **15**, 533 – 546.

COLLE, H.A. and REID, G.B., 1997, A framework for mental workload research and applications using formal measurement theory. *International Journal of Cognitive Ergonomics*, **1**, 303 – 313.

COOPER, G.E. and HARPER, R.P., 1969, *The Use of Pilot Rating in the Evaluation of Aircraft Handling Qualities* (Moffett Field, CA; NASA Ames Research Center). Report number NASA TN-D-5153.

CRONBACH, L.J., 1951, Coeffcient alpha and the structure of psychometric tests. *Psychometrika*, **31**, 93 – 96.

FIELD, E.J., 1995, Flying qualities of transport aircraft: precognitive or compensatory? PhD thesis. College of Aeronautics. Cranfield University. Cranfield, UK.

GAWRON, V.J., 2000, *Human Performance Measures Handbook* (Mahwah, NJ: Lawrence Erlbaum Associates).

HAIR, J.F., ANDERSON, R.E., TATHAM, R.L. and BLACK, W.C., 1998, *Multivariate Data Analysis*, 5th edition (Upper Saddle River, NJ: Prentice Hall).

HARPER, R.P. and COOPER, G.E., 1986, Handing qualities and pilot evaluation. Journal of Guidance. *Control and Dynamics*, **9**, 515 – 530.

HARRIS, D., 2000, The measurement of pilot opinion when assessing aircraft handling qualities. *Measurement and Control*, **33**, 239 – 243.

HARRIS, D., GAUTREY, J., PAYNE, K. and BAILEY, R., 2000, The Cranfield Aircraft Handling Qualities Rating Scale: a multidimensional approach to the assessment of aircraft handling qualities. *The Aeronautical Journal*, **104**, 191 – 198.

HARRIS, D., PAYNE, K. and GAUTREY, J., 1999. A multidimensional scale to assess aircraft handling qualities. In *Engineering Psychology and Cognitive Ergonomics*, volume three, D. Harris (Ed.), pp. 277 – 285 (Aldershot: Ashgate).

HART, S.G. and STAVELAND, L.E., 1988, Development of the NASA task load index TLX: Results of empirical and theoretical research. In *Human Mental Workload*, P.A. Hancock and N. Meshkati (Eds.), pp. 139 – 183 (Amsterdam: North-Holland).

NEWMAN, R.L. and GREELEY, K.W., 2001, *Cockpit Displays: Test and Evaluation.* (Aldershot: Ashgate).

PAYNE, K. and HARRIS, D., 2000, The development of a multi-dimensional aircraft handling qualities rating scale. *International Journal of Aviation Psychology*, **10**, 343 – 362.

PARKER' S, 2002, Online car reviews. Available online at: http://www.parkers.co.uk/choosing/ (accessed 11 March 2005).

ROBINSON, J.P., SHAVER, P.R. and WRIGHTSMAN, L.S., 1991, Criteria for scale selection and evaluation. In *Measures of Personality and Social Psychological Attitudes*, J.P. Robinson, P.R. Shaver and L.S. Wrightsman (Eds.), pp. 1 – 15 (San Diego, CA: Academic Press).

ROSCOE, A.H., 1984, Assessing pilot workload in flight. Flight test techniques. In *Proceedings of NATO Advisory Group for Aerospace Research and Development AGARD*, (Neuilly-sur-Seine: AGARD). Report number AGARD-CP-373.

SANO, S., 1982, Evaluation of motor vehicle handling. *International Journal of Vehicle Design*, **3**, 171 – 189.

SOCIETY OF AUTOMOTIVE ENGINEERS, 1985, *Subjective Rating Scale for Vehicle Handling*, (Warrendale, PA: Society of Automotive Engineers). Surface Vehicle Recommended Practice J1441, Issued 1985 – 06.

STANTON, N.A. and YOUNG, M.S., 1999a, Utility analysis in cognitive ergonomics. In *Engineering Psychology and Cognitive Ergonomics*, volume four, D. Harris (Ed.), pp. 411 – 418 (Aldershot: Ashgate).

STANTON, N.A. and YOUNG, M.S., 1999b, *A Guide to Methodology in Ergonomics* (London: Taylor and Francis).

TESTING TECHNOLOGY INTERNATIONAL, 1998, *Defining the Vehicle Development Process.* Vol. 1 (May 1998).

WALKER, G.H, STANTON, N.A. and YOUNG, M.S., 2001, An on-road investigation of vehicle feedback and its role in driver cognition: implications for cognitive ergonomics. *International Journal of Cognitive Ergonomics*, **5**, 421 – 444.

WHAT CAR?, 2001, Road test library. Available online at: http://www.whatcar.com/default.asp?a = roadtest (accessed 11 March 2005).

WILSON, D.J. and RILEY, D.R., 1989, Cooper-Harper rating variability. *Presented at AIAA Atmospheric Flight Mechanics Conference*, (Boston, MA: American Institute of Aeronautics and Astronautics). Paper number A1AA-89 3358.

Appendix 3:
Huddlestone and Harris (2007)

Huddlestone, J. & Harris, D. (2007). Using Grounded Theory techniques to develop models of aviation student performance. *Human Factors and Aerospace Safety, 6*, 357–368.

Commentary

Unlike the previous papers which were lengthy and highly statistical in their nature, this paper used the qualitative analysis of existing transcripts containing instructor comments about the performance of fast jet pilots and navigators.

In common with many papers using a qualitative analytical approach, one of the main challenges was to instil in the reader the confidence that the data were meticulously analysed and hence that the conclusions drawn were valid. To this end the emphasis in this paper was very much upon demonstrating that an appropriate analytical methodology had been rigorously employed (Grounded Theory) and that everyone was seeing the same things in the data (emphasis on inter-rater reliability). A second issue was in ensuring the clear interpretation of the data for the reader alongside the analysis process, which subsequently led to the model development and discussion of the results within a wider context. It had to be obvious that there was an audit trail back to the original raw, narrative data.

A further challenge was that as a result of space limitations (the edition of the journal in which the work was published was a special issue made up of the best papers delivered at the 2006 European Association for Aviation Psychology conference) all the qualitative data and its analysis had to be described in a relatively short paper (just less than 4,000 words). Usually, as a result of the nature of the data, qualitative analyses tend to be somewhat lengthy.

The 'Big Message'

A rich source of performance data exists in many training organisations which is not utilised to full effect but which can be used to develop useful models of trainee performance, allowing the structured analysis of student progression and teaching effectiveness.

The Story

- Narrative training data are collected in organisations which are not used to full effect.
- Communication, information interpretation and decision-making are key activities when crews in a number of aircraft are working together to achieve a common objective.
- Grounded Theory techniques are an appropriate method to develop a model of student performance from this archival, unstructured data which can be used for subsequent student evaluation.
- Using this approach a relatively small number of reliable categories could be elicited from the analysis of post-sortie, instructor assessments to produce a comprehensive description of student performance.
- These categories were further developed into a student performance model with components within it recognisable from earlier related research, hence giving evidence of its validity.
- This approach could be utilised to develop models from extant, unstructured data to describe trainee performance in other areas.

Human Factors and Aerospace Safety 6(4), 371-382
© 2007, Ashgate Publishing

Using Grounded Theory techniques to develop models of aviation student performance

John Huddlestone and Don Harris
Cranfield University, UK

Abstract

A frequently encountered issue in both aviation training and aviation training research is that of student performance modelling. A potentially rich source of student performance data exists in many if not all aviation training organisations in the form of narrative reports written after flying or simulator sorties. This paper describes how grounded theory technique, developed by Glaser and Strauss (1967) was used to perform sampling and analysis of such data to develop a model of student performance. The application of the technique is illustrated with a case study from the military flying training domain.

Introduction

A frequently encountered issue in aviation training research is that of student performance modelling. A potentially rich source of student performance data exists in many if not all aviation training organisations in the form of narrative reports written after flying or simulator sorties. The challenge is in finding a suitable way to sample and then analyse the data in these reports to produce a meaningful model.

Grounded theory technique, however, developed by Glaser and Strauss (1967), provides a rigorous and structured method to perform sampling and analysis of

Correspondence: John Huddlestone, Flight Operations Research Cenhtre of Excellence, Department of Human Factors, School of Engineering, Cranfield University, Cranfield, Bedford, MK43 0AL, UK or e-mail j.huddlestone@cranfield.ac.uk

such unstructured data. Strauss and Corbin (1990) defined grounded theory as follows:

'A grounded theory is one that is inductively derived from the study of the phenomenon it represents. That is, it is discovered, developed and provisionally verified through systematic data collection and analysis of data pertaining to that phenomenon...One does not begin with a theory then prove it. Rather, one begins with an area of study and what is relevant to that area is then allowed to emerge.' (p. 23)

The application of this technique is illustrated with a case study from the military flying training domain.

Background to the study

A performance model was required for fighter crews undergoing initial training in pairs combat techniques on the Royal Air Force Panavia Tornado F3 Operational Conversion Unit (OCU). The main course run by the OCU was designed to train fighter crews (pilots and navigators) in the operation of the Tornado F3 in the air combat role. The final phase of the course focussed on pairs tactics. The use of a pair of visually supporting aircraft has formed the basis of virtually all air combat tactics since aircraft have been used in combat (Tornado F3 Training Course Pairs brief, 1998). Leadership of a pair of aircraft is heavily dependent on effective communication between the aircraft crews, as they have to build and maintain situation awareness whilst engaged in a fast moving, three-dimensional fight and make appropriate tactical decisions.

The results of a task analysis of Beyond Visual Range (BVR) intercepts conducted by Houk, Whitaker and Kendall (1993) were used by Waag and Houk (1994) to derive a set of behavioural indicators suitable for evaluating proficiency in air combat in day to day squadron training. They identified communication, information interpretation and decision-making as key activities. Bell and Lyon (2000) reported that communication was one of the most highly rated elements contributory to good situation awareness, based on a survey of mission ready McDonnell-Douglas F-15C fighter pilots. A high level of situation awareness was also identified as a critical component for effective decision making by Endsley and Bolstad, (1994). Furthermore, it has been argued that the process of situation assessment (encompassing such activities as communication and information interpretation) is a fundamental precursor of situation awareness, which is itself the precursor for all aspects of decision-making (Nobel, 1993; Prince and Salas, 1997). These propositions put forward in these studies are also consistent with the model of situation awareness and its position in the decision-making, action-taking loop proposed by Endsley (1995).

Using Grounded Theory to develop models of student performance 373

Anecdotal evidence from the instructors on the OCU suggested that students were weak at communication during this final phase of training. They attributed this to the unsuitability of the ground training device in use, which only took a single crew. A multi-player simulator was to be evaluated to determine if it produced positive transfer of training to the airborne environment for pairs training. The empirical evidence, in the form of archived narrative reports of past student performance in the pairs phase, was investigated in order to develop a student performance model. This model was required both to validate the anecdotal evidence from the instructors about students' performance, and to subsequently inform the development of suitable behavioural indicators for the evaluation of the multi-player simulator.

Method

Overview

The approach taken to applying grounded theory to this study was based on the procedures and techniques suggested by Strauss and Corbin (1990). Figure 1 shows the sequence of steps that were followed.

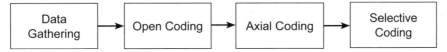

Figure 1 **Grounded theory application steps (from Strauss and Corbin, 1990).**

The following sections describe the nature of each of these steps and the methods that were employed in implementing them.

Data gathering

The data were collected from pairs-phase sortie reports using what Strauss and Corbin (1990) describe as a line-by-line analysis technique. Reports were selected from courses completing the pairs phase during a three-month period. Each report was broken down into individual sentences or small groups of sentences that referred to a single observation about an aspect of performance. Three sampling criteria were employed at this stage. The first element was concerned with the selection of observations directly relevant to practice intercepts (sortie reports covered the whole of the sortie from starting up the aircraft and taxiing out to the recovery back to base). The second element of sampling was aimed at achieving a

374 *John Huddlestone and Don Harris*

balanced view of both pilot and navigator assessment issues.   To achieve a representative sample covering all performance aspects of the practice intercepts approximately 100 comments for each crew member were selected from the sortie reports. The final sampling criterion was to select from as wide a range of instructors who had written the assessments as possible.

Open coding

Open coding is defined by Strauss and Corbin (1990) as:

> *'The process of breaking down, examining, comparing, conceptualising and categorising data.'*

The tabulated comments produced by the line-by-line analysis during data gathering were analysed to identify categories into which they could be grouped. This analysis was conducted using the constant comparison technique, as described by Partington (2002), whereby each data item was compared with preceding items to see if it described the same phenomenon as one of the items and therefore could be allocated to an existing category, or if a new category needed to be developed. The application of this to the set of instructor comments by the first rater yielded an initial set of coding categories. The comments were then re-coded using these categories by a second rater, to check for inter-rater reliability. Differences in coding were discussed and resolved to produce an agreed list of categories.

Axial coding

Strauss and Corbin (1990) define axial coding as:

> *'A set of procedures whereby data are put back together in new ways after open coding, by making connections between categories.'*

The output from this stage was a set of higher order categories describing student performance along with a description of the nature of the connection between the lower order categories that they contained. This required a detailed re-evaluation of the comments within each category.

Selective coding

Selective coding is defined by Strauss and Corbin (1990) as:

> *'The process of selecting the core category* [and] *systematically relating it to other categories.'*

Using Grounded Theory to develop models of student performance 375

They also point out that this process is essentially similar to axial coding, but is conducted at a higher level of abstraction. The core category is the over-arching phenomenon or concept that links each of the categories or phenomena that are developed during axial coding. After the core crew performance category had been identified, the links between the categories identified during axial coding were revealed through further analysis of the comments in each category. Once all the categories had been linked together to form a complete model, a narrative description was developed.

Performance model development

Data

A total of 200 performance comments were collected from 46 sortie reports. There was an even split of navigator and pilot student reports. Twenty-eight (80% of the instructor population) completed reports that were used in the sample.

Open coding

The open coding process conducted by the first investigator yielded eight initial coding categories, shown in the first column of table 1. During the open coding process it was noted that the performance comments could be categorised as being either positive or negative comments. Each comment was annotated accordingly. Columns two and three of table 1 show the frequencies of positive and negative comments or each of the initial coding categories that were identified. Column four shows the total number of comments allocated to each category. A striking feature of the data was that 89 out of 200 (45%) of the performance statements were related to communications.

When the second investigator then re-coded the performance statements, using the categories identified by the first investigator, agreement was achieved for 70% (140) of the performance statements.

During the subsequent discussion between the investigators to resolve differences of opinion about coding categories, the most problematic category was 'decision-making'. In 40 of the 60 instances where there was a disagreement, the issue was whether a statement should be categorised as 'decision-making' or whether it should be categorised as 'weapons employment', 'communications' or 'tactics'. In the subsequent discussions it was identified that decision-making was taking place in a range of contexts and that new categories may have been appropriate for 'weapons decision-making' and 'communications decision-making'. The following statement about 'communications' was one of a number of statements illustrated that a 'communications decision-making' category was justified:

376 *John Huddlestone and Don Harris*

'If things are worth saying to your pilot, they could be worth transmitting.'

This statement suggested that, on some occasions at least, the student should have decided to communicate with the other aircraft and didn't. Similarly, statements were found that supported the creation of a 'weapons decision-making' category.

Table 1   Initial open coding categories and positive, negative and total comment frequencies for each category.

Initial Open Coding Catgory	Frequency of Positive Comments	Frequency of Negative Comments	Total Number of Comments
Communications	26	63	89
Weapons Employment	5	21	26
Situation awareness	6	20	26
Tactics	7	17	24
Decision Making	6	12	18
RHWR Awareness	2	4	6
Use of Chaff and Flare	3	4	7
Leadership	2	2	4
Total	**57**	**143**	**200**

Having identified that the communications and weapons employment categories could in part be decomposed in to related decision-making sub-categories, the view was taken that theoretical sensitivity had strongly influenced the development of the initial open coding categories. 'Communications' was expected as a category given that it featured centrally in the research question being addressed and this could account for the underlying categories being overlooked initially. The constant comparison technique was applied again to the statements in the 'communications' and 'weapons employment' categories to see if data supported this view.

In the case of the remaining 'weapons employment' statements it was found that they were all concerned with the operation of the weapons system, hence the 'weapons system operation' category was introduced. The 'communications' category statements were found to relate to the passing of information in different contexts. Further application of the constant comparison technique to these statements yielded sub-categories for 'passing tactical situation and actions', 'giving tactical orders' and 'passing missile shot information'.

As with 'communications', 'situation awareness' had been expected as a category and comments about it were often placed under a heading of 'situation awareness' in the narrative reports by the instructors. Re-analysis of the set of

Using Grounded Theory to develop models of student performance 377

comments made about 'situation awareness' revealed that the instructors viewed 'situation awareness' as a set of processes (q.v. 'situation assessment'). The need for students to monitor displays and communications as sources of information for situation awareness is highlighted in comments such as:

> '*At times his situational awareness was not as high as it should have been, he must build it through listening and monitoring the tactical display.*'

Other statements implied that 'situation awareness' also involves processing this information in some way in order to make judgements. For example, the following statement indicates that evaluation of the bandits' (enemy) actions and anticipation of their possible intentions were considered essential activities:

> '*... situational awareness – don't just say what the bandits are doing – try to draw some conclusions about their behaviour and our RHWR* [Radar Homing and Warning Receiver] *indications*'

Based on this analysis, the 'situation awareness' category was decomposed into 'monitor', 'evaluate' and 'anticipate' sub-categories (effectively the 'situation assessment' processes underpinning 'situation awareness').
Finally, it was identified that all of the statements categorised as either 'decision-making' or 'tactics' by the first investigator could be accurately described as referring to tactical decision-making. Consequently, the 'tactics' and 'decision-making' categories were merged to create a single 'tactical decision-making' category.

Table 2   Initial and final open coding categories.

Initial categories	Final Categories			
Communications	Comms decision making	Pass tactical situation/ actions	Give tactical orders	Pass missile shot information
Weapon Employment	Weapons decision making		Weapon system operation	
Situation awareness	Monitor	Evaluate		Anticipate
Tactics	Tactical Decision Making			
Decision making				
RHWR awareness	RHWR awareness			
Use of chaff and flates	Use of Chaff and flares			
Leadership	Leadership			

378 *John Huddlestone and Don Harris*

The remaining categories were found not to subdivide. The final agreed set of categories is shown in the second column of table 2.

Axial coding

During the open coding stage, 'decision-making' was identified as a central component of 'tactics', 'weapons employment' and 'communications'. The instructor comments in the remaining categories were reviewed to identify if it occurred elsewhere. All the statements in the 'use of chaff and flares' category were also found to refer to decision-making. For example:

> '[Use] *flares at the correct range otherwise, as you saw, you will get a Fox 2* [infra-red guided missile] *in the face'*

Clearly, a poor decision about flares usage could have fatal consequences. The operation of chaff and flares is a matter of a button press so it is not surprising that there were no statements about chaff and flares operation. From this analysis, 'decision-making' was identified as a high-level category which included the 'communications', 'weapons' and 'tactical decision-making' categories along with the 'use of chaff and flares' category.

During open coding, 'situation awareness' was decomposed into the underpinning process activities of 'monitoring' the environment, 'evaluating' this information and 'anticipating' future events (situation assessment). The following comment suggested that the 'RHWR awareness' could also be regarded as an aspect of monitoring the environment:

> '*He is not monitoring his RHWR, he made no mention of threat indications – he must bring it into his scan'*

Therefore, this category was subsumed within the 'monitoring' category. 'Situation awareness' was reintroduced but as a high level category containing the previously mentioned sub-categories describing the processes of 'monitor', 'evaluate' and 'anticipate'.

The open coding categories of 'weapons system operation', 'passing tactical situation/actions', 'giving tactical orders' and 'passing missile shot information' could be construed as actions taken in response to decisions. Thus, the 'actions' category was created containing these sub-categories. 'Leadership', the remaining open coding category, was not considered to fit into any other category.

Selective coding

The key to the selective coding stage and the subsequent development of an overall model was the identification of the core category. This was required in

Using Grounded Theory to develop models of student performance 379

order to facilitate the integration of the 'decision-making', 'action' and 'situation awareness' and 'leadership' categories identified in the axial coding stage. As a result of re-examining the full set of statements, the following statement was identified which pointed to a potential core category:

> '... he must be more missile launch success zone aware, he was generally trying to do all things rather than the main priority – killing the bandit.'

The significance of this statement is that it identified the primary goal of all the activities engaged in during the practice intercepts, which was killing the enemy.

The next stage was to identify the relationships between high-level categories. The connection between 'decision-making' and 'actions' was simple; actions were taken in response to decisions made. The connection between 'situation awareness' and these categories had to be established. A clear insight into the purpose of situation awareness, namely that it informs decision-making, was given by the statement:

> '... he lost situational awareness momentarily and shot at his wingman'

In this instance poor situation awareness resulted in an erroneous weapon employment decision. The question then arose as to what happened once an action had been taken? A number of comments pointed to the iterative nature of the process such as:

> 'If a missile shot does not come off shoot again or reposition'

The inference drawn from this comment was that following an action the situation had to be monitored and re-evaluated so that subsequent decisions could be taken and actioned.

Further analysis of the leadership statements showed that effective leadership of the pair during the intercept required good situation awareness and sound decision-making and was implemented though effective communication.

Discussion

Figure 2 shows the complete performance model, with situation awareness, decision-making and action categories broken down into their sub-components. The central feature of the model is the constant repetition of development of 'situation awareness' leading to 'decision-making' and subsequent 'actions'. 'Communication' is prominent, both as an aspect of decision-making and as a required action following decision-making. These features of the model were consistent with the findings of Waag and Houk (1994) in their development of a

380 *John Huddlestone and Don Harris*

set of behavioural indicators suitable for evaluating proficiency in air combat in day-to-day F-15 squadron training. They identified communication, information interpretation and decision-making as key activities. This point is reinforced by Bell and Lyon (2000) who reported that communication was one of the most highly rated elements contributory to good situation awareness, based on a survey of mission ready F-15C fighter pilots. The monitor-evaluate-anticipate model of situation awareness was found to be directly equivalent to the perception-comprehension-projection model of situation awareness proposed by Endsley (1995). Therefore, the student performance model that was developed was considered to be highly consistent with the published models of air combat and situation awareness. The OCU instructors reviewed the model and agreed that it was consistent with their perception of the air combat task and accurately captured student weaknesses in performance.

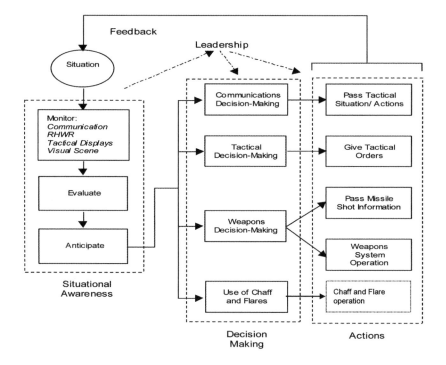

Figure 2   Detailed student performance model

Using Grounded Theory to develop models of student performance 381

One of the instructors responsible for teaching sessions on the use of the radar during the basic radar phase at the start of the course commented that the model developed was very close to the model which he taught, shown in figure 3.

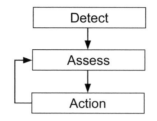

Figure 3   Instructor model for radar use.

In this model the detect component referred to the initial search for target aircraft. Once a target had be spotted, the assess stage covered the monitor, evaluate and anticipate components of situation awareness. The action component referred to the decision-making action sequence. The parallel between this model and that developed through the grounded theory study adds support to the argument that the model produced was valid. It also supported the view that, by definition the items that were considered important for comment reflected the training priorities of the instructors and as such represented the model of air combat that they were teaching.

The grounded theory techniques applied during this study provided a powerful toolset for the analysis of the sortie report data, enabling a detailed model of student performance during the pairs phase to be built.

Returning to the initial objectives of the study the principal reason for developing the model was to validate the instructors' view, expressed anecdotally, that communications were a major area of weakness in student performance in the pairs phase. The analysis of the number of comments in each of the categories identified in developing the model showed that 45% of the comments were made about communications and that the majority of those were negative, validating the instructors' perception that students' performances were weak in the area of communications. Weaknesses in the areas of 'situation awareness' and 'decision-making' were also exposed. The presence of both positive and negative comments in each of the open coding categories that were developed was interpreted as indicating that these aspects of performance were considered to be important by the instructors as they were choosing to comment on performance in these areas regardless of whether or not a student had made a mistake. Indeed, the grounded theory analysis clearly identified eight basic areas (and nine further sub-areas) for meaningful assessment of student performance in the Tornado F3 pairs phase. Therefore, the model provided a sound basis upon which to determine candidate measures for a subsequent training effectiveness trial.

From a methodological perspective, the grounded theory techniques applied during this study facilitated rigorous analysis of the sortie report data, enabling a detailed model of student performance during the pairs phase to be built. Investigator triangulation was found to be essential in order to reduce bias caused by theoretical sensitivity. It is suggested that Grounded Theory is a technique that is very well suited to the analysis of qualitative performance data in the form of narrative reports which are commonly used in the aviation training domain.

References

Bell H.H. and Lyon, D.R. (2000). Using Observer Ratings to Assess Situational Awareness. In, M.R. Endsley and D.J. Garland (Eds.) *Situation Awareness Analysis and Measurement.* New Jersey: Laurence Erlbaum Associates (pp. 129-146).

Endsley M.R. (1995). Toward a Theory of Situation Awareness in Dynamic Systems. *Human Factors, 37,* 32-64.

Endsley, M.R. and Bolstad, C.A. (1994). Individual Differences in Pilot Situation Awareness. *International Journal of Aviation Psychology, 4,* 241-264.

Glaser B. and Strauss A. (1967). *The Discovery of Grounded Theory.* Chicago: Aldine.

Noble, D. (1993). A Model to Support Development of Situation Assessment Aids. In, G.A. Klein, J. Orasanu, R. Calderwood, and C.E. Zsambok (Eds), *Decision Making in Action: Models and Methods.* Norwood, New Jersey: Ablex. (pp. 287-305).

Partington, D. (2002). Grounded Theory. In, D. Partington (Ed) *Essential Skills for Management Research.* London: Academic press (pp. 136-157).

Prince, C. and Salas, E. (1997). Situation Assessment for Routine Flight and Decision Making. *International Journal of Cognitive Ergonomics, 1,* 315-324.

Strauss A. and Corbin, J. (1990) *Basics of Qualitative Research: Grounded Theory Procedures and Techniques.* London: Sage.

Waag W.L. and Houk M.R. (1994). Tools for Assessing Situational Awareness in an Operational Fighter Environment. *Aviation, Space and Environmental Medicine, 64 (5, Suppl)* A13-A19.

Appendix 4:
Demagalski, Harris and Gautrey (2002)

Demagalski, J.M., Harris, D. & Gautrey, J.E. (2002). Flight control using only engine thrust: development of an emergency display system. *Human Factors and Aerospace Safety*, 2, 173–192.

Commentary

This final paper is what may almost be classified as a 'traditional', experimental paper. However, in this case the experimental trials in question were undertaken in the ecologically-valid setting of a flight simulator. The initial stimulus for the research stemmed from the crash landing of United Airlines flight 232 at Sioux City. After complete loss of its flight control system the aircraft was flown manually by the pilots using only differential throttle. Although many people were killed in the accident that followed the skill of the pilots also saved a great number. It also demonstrated that control of a large aircraft could be exercised using only engine power. A great deal of research was subsequently undertaken in the US developing fully automatic, engine-only emergency control systems, however these were extremely expensive and difficult to certificate. Furthermore, a flight control system total failure is extremely rare. The system developed and trialled in this study used only an emergency display system to guide the pilots. It was designed to provide 99% of the 'bang' of the fully automated emergency flight control systems for only 1% of the 'buck'.

The challenges in writing this paper were that not only did it involve complex trials, the design and operation of the emergency display system also had to be explained, as did the flight scenarios in which it was tested. Furthermore, there was little extant Human Factors research to provide the theoretical background for the Introduction for the paper, which also left little to be said in the Discussion, making the research challenging to place within a scientific context. Just to make things a little more difficult, a colour display system also had to be presented in a journal published in black and white!

The 'Big Message'

With a little bit of help from a display system that we have designed, in an emergency you can fly an airliner using just differential thrust from the engines and execute a technically-survivable landing.

The Story

- In the rare event of the total failure of the flight control system, skilled pilots have demonstrated that they can control an aircraft using just differential engine thrust.
- Fully automatic, emergency flight control systems have been developed but these are very expensive and difficult to certificate.
- A low cost emergency flight display system to aid pilots in controlling the aircraft has been developed which was effective even when pilots had no previous training in using the system.
- This display was tested in an ecologically-valid setting using a representative range of flight scenarios using professional pilots.
- The emergency display system worked well compared to the control condition (no display) allowing pilots to execute technically survivable landings using thrust only control.
- Further developments for the display system have been identified which should enhance its performance even further.

Human Factors and Aerospace Safety 2(2), 173-192
© 2002, Ashgate Publishing

Flight control using only engine thrust: development of an emergency display system

Jason M. Demagalski, Don Harris and James E. Gautrey
Cranfield University, United Kingdom

Abstract

The total failure of an aircraft's flight control system is an extremely rare event but in such cases pilots have demonstrated that some control may be regained by the judicious use of engine thrust. Complex, reversionary autoflight control systems have been developed using this technique. This paper describes the development and testing of an emergency display system, designed to use only existing on-board sensors and displays, to allow the crew to assume manual control and effect an emergency landing in such an event. Initial tests in an engineering flight simulator using a sample of line pilots suggest that when using this simple display system they were able to recover control of the aircraft and navigate it with sufficient accuracy to effect a technically-survivable emergency landing.

Introduction

The chances of losing the ability to control a modern commercial aircraft as a result of a failure of its flight control system (FCS) are extremely remote. Airworthiness regulations (such as FAR/JAR 25) require that the likelihood of such a failure should be less than one in every billion (1×10^{-9}) flight hours. However, such failures do happen and almost always with catastrophic consequences. Perhaps the most famous instance of an event of this type was the accident involving United Airlines flight 232 (NTSB, 1990). In this accident the number two engine in this McDonnell Douglas DC10 suffered a catastrophic failure of the first stage fan blade. Debris was ejected from the engine nacelle severing all the hydraulic lines to the control surfaces on the tail of the aircraft. Subsequently all hydraulic pressure

Correspondence: Jason Demagalski, Human Factors Group, Cranfield University, Cranfield, Bedford, MK43 0AL, United Kingdom, or j.demagalski@cranfield.ac.uk

174 *Jason M. Demagalski, Don Harris and James E. Gantry*

was lost and the aircraft became almost uncontrollable in roll, pitch and yaw using the conventional flight controls. However, the crew managed to recover some control by the use of differential and symmetrical thrust changes on the remaining two serviceable engines. By using this technique the crew were almost able to effect a successful emergency landing at Sioux City airport. Unfortunately, just before touchdown the aircraft developed a very high sink rate and a slight roll. The aircraft broke up as it hit the runway with a sink rate of 1,620 feet per minute. One hundred and eleven passengers and crew were killed however 185 people survived as a result of the skill and initiative of the pilots.

While such an occurrence is extremely rare there have been other instances where control has been lost as a result of a full or partial failure of the FCS. Four years prior to the United Airlines flight 232 accident a Boeing 747 crashed into a mountainside near Tokyo killing 520 passengers and crew after the FCS lost all hydraulic pressure following the failure of the aft pressure bulkhead. In 1973 a US Air Force Lockheed C5A crashed in Saigon killing 178 people. The crew almost maintained control of the aircraft using aileron and differential thrust when they lost hydraulic pressure to the rudder, elevator and flaps, but finally lost control 1½ miles from the runway threshold. The following year a McDonnell Douglas DC10 crashed just outside Paris killing 346 people after an explosive decompression following the failure of the aft cargo door. The collapse of the aircraft's floor following the decompression damaged the hydraulic lines and the tail of the aircraft, compromising the pilots' control. In other cases of total hydraulic failure though (e.g. Delta Lockheed L-1011, San Diego, 1979; American Airlines Flight 96, Windsor Ontario, 1972) the pilots successfully regained control of the aircraft using the aircraft's throttles as a reversionary FCS.

However, it is not just the loss of the FCS that can compromise the control of an aircraft. There have been several incidents of in-flight icing jamming the aircraft's control surfaces (e.g. AllStar Airlines DC9, 1985; Northwest Airlines Boeing 757, 1986; Alaska Airlines SA 227, 1995; US Airways Express DHC-8, 2000). Fortunately in all these cases control was regained after the ice melted following descent to a lower altitude effected by reducing engine power. Differential and symmetrical throttling of the engines was again the principal method by which control was regained.

Blezad (1996) describes the problem faced by the pilot of the accurate control of flight path using changes in thrust alone. 'The fundamental problem is that the pilot with the throttles alone cannot easily and predictably maneuver an aircraft. Although the control power is often sufficient to fly the aircraft, the long time constants and couplings between dynamic modes make pilot control uncertain and precarious for tasks such as landing'.

The greatest problem faced by the pilots in such a situation is posed by the phugoid mode. The phugoid is a long-period pitch oscillation that occurs when an aircraft's speed is changed from its trim speed. The phugoid is suppressed whenever the pilot holds an attitude, altitude or airspeed. Normally it is counteracted through control inputs to the elevator, however this is not possible in the case of an FCS failure. Furthermore, in the event of an FCS failure the trim airspeed is set to that at the time of the failure without any possibility of changing it. Speed and pitching interactions become inseparably cross-coupled. For example, if the pilot attempts to descend by rapidly reducing power, the aircraft will initially enter a descent however it will simultaneously also begin to increase speed. As airspeed increases lift increases which eventually has the effect of arresting the aircraft's descent and initiates a nose-up, climbing attitude. As the aircraft begins to climb its airspeed decays, until it reduces to such a point that the nose pitches down again and the whole oscillatory cycle begins once more. In a large commercial aircraft this oscillatory period is usually of the order of one minute. The oscillations eventually damp themselves out over a number of cycles as airspeed returns to the trimmed airspeed (trimmed angle of attack).

A similar problem faces the pilot in the lateral control axis. Many aircraft exhibit a coupled mode between yaw and roll, called Dutch Roll in which the nose of the aircraft traces an elliptical path. This causes difficulties for the pilot flying by throttle control alone (as turns must be initiated using differential levels of thrust which initially induces a yawing motion, followed by the aircraft rolling into the turn) but does not compromise control to the same extent as the phugoid mode (Burcham, Fullerton, Gilyard, Wolf and Stewart, 1991).

Pilots have little control over the phugoid motion once it commences. The DC10 that crashed at Sioux City had an excessive rate of descent at touchdown because the crash landing occurred at the wrong stage of the oscillatory cycle. Judicious use of small thrust inputs at the appropriate point in the cycle can help to damp the phugoid, however it is difficult for the pilot to predict exactly when this is. Furthermore, it can also, take between 20-40 seconds for a commanded change in thrust to have an effect on the aircraft's behaviour. This is difficult to predict as the time required for a modern turbofan engine to spool-up depends upon the amount of thrust being delivered prior to the thrust lever input. Most engines respond faster at higher levels of thrust.

These control delays are problematical from a human control point of view. It is well documented that large control lags induce poor human control performance (e.g. Wickens, 1992). It is very difficult to predict the magnitude (and timing) of a throttle input to produce the desired aircraft response. Human controllers are also very poor at predicting the outcome of a control response when it also requires the prediction of the future position, velocity and acceleration of the system they are trying to control (see McRuer et al., 1968; Wickens, 1992) as is the case when predicting flight path in response to changes in engine thrust while also taking into account the phugoidal motions of the aircraft. Clearly, the pilot needs considerable assistance when using thrust as the primary means of control.

After the Sioux City accident NASA (National Aeronautics and Space Administration) instigated a programme of work to investigate the controllability of multi-engined aircraft using throttles alone (Burcham et al, 1991). In a series of flight tests it was found that all aircraft were controllable using throttles alone, however precise control was difficult as a consequence of the lags associated with engine response times and the phugoid mode. It was also observed that pilot proficiency improved rapidly with practise. Repeatable landings were possible with certain aircraft types (specifically the Boeing 720 and the McDonnell-Douglas F-15) using only the throttles.

Having demonstrated that it was possible to control a multi-engined aircraft using thrust alone it was decided to pursue the use of differential engine thrust as an emergency FCS but not by using a 'pilot in the loop' system. Instead a closed-loop, autoflight control system was preferred (Maine et al., 1994). The NASA PCA (propulsion controlled aircraft) programme initially used a McDonnell Douglas (Boeing) MD-11 aircraft as its test-bed and latterly a Boeing 747-400 flight simulator. Longitudinal flight path changes were effected by parallel changes in engine thrust and lateral flight path changes by asymmetric changes. The PCA software controlled each engine's EPR (engine pressure ratio) directly via the FADEC (full-authority digital engine control). The throttle levers were bypassed and their servo back-drive was non-operational when the aircraft was flown in this mode. The PCA software demonstrated that it was possible to control the aircraft within ±0.5° in pitch and roll and effect a safe emergency landing with a rate of descent at touchdown in the order of 4-500 feet per minute (Bull, Mah, Hardy, Sullivan, Jones, Williams, Soukup and Winters, 1997).

The greatest drawback to the implementation of the PCA concept as an emergency FCS was that it required extensive changes to the flight-critical FADEC software and hence required re-certification. It also invalidated the engine manufacturer's warranties. As a result it was thought that it was unlikely to be accepted by the airlines, regulatory authorities or manufactures. In response, NASA developed the 'PCA-Lite' system, which effected control via the autothrottle and thrust-trimming systems instead of the FADEC, thereby avoiding costly re-certification. Burcham, Maine and Burken (1998) demonstrated a similar level of performance for this system as for the original system.

NASA also tested a 'PCA-Ultralite' system that only controlled the aircraft in pitch. Pilots were responsible for controlling lateral flight path by differentially manipulating the throttles as required (Burcham et al., 1998). Pilots that already had experience of the PCA system coped well but those with no experience tended to over-control the aircraft resulting in unstabilised approaches and landings with a high vertical speed component. Burcham, Maine, Kaneshige and Bull (1999) suggested that a flight deck display system to guide pilots' throttle inputs was necessary.

Design concept and objectives for the controlled flight by throttles system

The design criteria for the emergency display system for controlled flight by throttles developed in this research was that it should allow the crew to navigate the aircraft in both axes and effect an emergency landing with a vertical speed component on touchdown within survivable limits. As the system was designed as a standby system to be used only in the extremely rare event of a total FCS failure, it was designed to use only existing on-board sensors with little or no modification to the hardware on the aircraft. It was designed to be an additional software module hosted on the flight management computer (FMC). To avoid safety critical software certification issues the system was designed to not have any command authority over the engine thrust management software. Furthermore, as the system would only be used on extremely rare occasions, it had to be usable with little more than a simple briefing and a bare minimum of training.

As the control cross-coupling induced by the failure of the FCS means it is impossible to separate the control of speed from pitch, it was decided that the display of predicted flight path would be of limited utility as the pilot would be able to effect little direct control over it. Although the primary aim of flying is usually the accurate control of flight path, when exercising control using throttles alone the display of actual and required trajectory is of limited utility. As a result the throttle-controlled emergency flight control system was implemented using a command type display, similar in concept to the flight director bars on the primary flight display. The display system commanded the individual throttle positions required to effect turns, climbs or descents. A further objective of implementing the system in this manner was that instead of attempting to control and/or predict the phugoidal motion of the aircraft in the pitch axis and Dutch roll in the roll/yaw axes, it was decided that it would be better simply to avoid exciting these modes.

As part of the display design process it was established that for the Boeing 747-200 (on which the initial display concept was developed and tested) excitation of the phugoid mode was less likely if vertical flight path angle (FPA) was limited to less than $2.5°$. Similarly, it was established that if roll angles were limited to less than $10°$ this was less likely to either excite the phugoid mode or induce Dutch Roll.

The control of airspeed independent of flight path was not possible when controlling the trajectory of the aeroplane by the use of throttles alone. With an aircraft that exhibits speed stability, after any disturbing input the aircraft will ultimately return to the trimmed airspeed at the time of the FCS failure, irrespective of throttle position or attitude. As a result, no effort was made to control airspeed.

Design and operation of the emergency flight control display system

The display format was derived after a series of walk-through, talk-through task analyses and more formal task analyses undertaken in a full-flight simulator. This was followed by a series of design iterations and evaluations in a full-scale engineering flight simulator by a qualified test pilot. The final display system

178 *Jason M. Demagalski, Don Harris and James E. Gantry*

developed is illustrated in figures 1, 2a and 2b. To manage the extra functionality an additional mode was incorporated onto the mode control panel (MCP).

When the emergency, controlled flight by throttles mode was activated the first of three sub-modes was automatically initiated. This was the 'recovery mode' (the default mode). The other modes were the 'en-route' mode and the 'approach and landing' mode.

Overall concept

The emergency controlled flight by throttles display system was comprised of three basic elements: a throttle position, command-type display; a display of instantaneous FPA, and in the en-route and approach and landing modes, a profile predictor display of vertical flight path. It should be noted that the navigation display in most 'glass cockpit' aircraft already contains a predictive display of final heading when executing turning manoeuvres. The emergency flight control display system was implemented on the primary engine display screen in the centre of the flight deck above the throttle quadrant to maintain spatial and proximal control/display compatibility (Andre and Wickens, 1990).

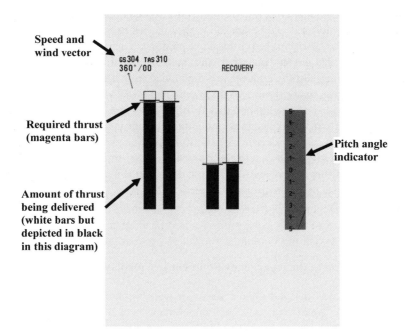

Figure 1   Emergency flight control display system in recovery mode

The basic operation that the display system was as follows. Once this emergency mode was activated, if a new heading was required this was entered into the heading window of the MCP. The computer then calculated the required thrust settings for each engine to initiate a turn onto the required heading while keeping the bank angle to less than 10° and simultaneously maintaining the aircraft in level flight. The required thrust settings were then indicated as magenta bars on the throttle position command display (see figure 1). On this display each engine was represented by a hollow vertical column. These columns were arranged to correspond with the layout of the aircraft's engines. Within each column was a white bar that represented the amount of thrust being delivered (as a percentage of maximum thrust available). The pilot's task was to adjust the amount of thrust being delivered on each engine to correspond to the required amount of thrust as indicated by the position of the magenta command bars. A degree of 'lead' was built into the system to aid the pilot when rolling out of the turn.

In the pitch axis the pilot entered the required FPA into the MCP window. These data were used to generate the required thrust settings to achieve the required FPA. When the vertical flight path profile display was active (except when in the recovery mode – see later) a magenta line representing the resulting predicted trajectory was displayed, along with a white line emanating from the nose of the aircraft indicating the instantaneous FPA (see figure 2a).

In approach mode, the magenta line displayed the glideslope required for landing (see figure 2b). To avoid exciting the phugoid mode during the final stages of approach and landing and to ease the transition from level flight, a 2° approach path was employed, rather than the conventional 3° approach. In this mode the magenta line on the vertical situation display was a graphical representation of the recommended 2° glideslope (not to scale). A scale along the left-hand side of the display indicated current height above ground level (AGL) and a scale along the bottom of the display indicated the distance (in miles) to go to the runway threshold. With respect to the lateral navigation aspect of the approach task, a conventional localiser-type of display was employed.

The emergency display system was completed by the inclusion of a digital display of ground speed (GS); true air speed (TAS); wind speed (all measured in knots) and strength (including a diagrammatic representation of the wind relative to the aircraft); time to the next waypoint; required heading to the next waypoint; and DME (distance measuring equipment) distance (in nautical miles) to the next waypoint. The display mode was displayed in the top, centre of the screen.

Display logic

Recovery mode   This was the default mode once the emergency system was activated. The objective of this mode was simply to recover the aircraft back into controlled straight and level flight as quickly as possible while losing as little

180 *Jason M. Demagalski, Don Harris and James E. Gantry*

altitude as possible. The magenta bars commanding the required throttle positions were programmed simply to achieve this aim.

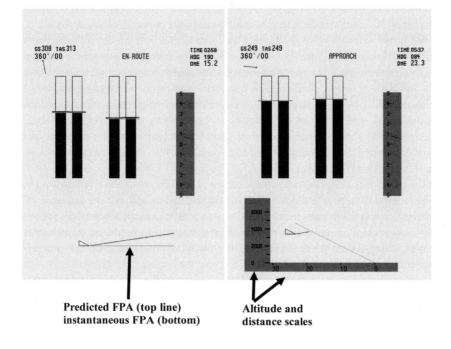

Predicted FPA (top line)   **Altitude and**
instantaneous FPA (bottom)   **distance scales**

Figure 2   Emergency flight control display system in en-route mode (figure 2a, left) and (approach and landing mode figure 2b, right)

En-route mode   After the aircraft had been recovered into straight and level flight the en-route mode was selected to steer the aircraft to the point at the top of final approach. In this mode roll angle was restricted to a maximum of 10° and for vertical manoeuvres pilots were advised not to use an FPA of greater than 2.5°.

Approach mode   The objective of the approach mode was not to enable the aircraft to make a perfect landing on the centreline of the runway. The aim was to enable the pilot to make a controlled, emergency landing on (or near) the runway where the airport's emergency services would be close at hand. At Sioux City the aircraft touched down with a 1,600 feet per minute (8.15 ms⁻¹) vertical speed component while on the downward phase of the phugoid. The undercarriage on most modern airliners will collapse if the aircraft lands with a vertical speed component in excess of about 800 feet per minute (4.04 ms⁻¹). At up to 1,000 feet per minute (5.05 ms^{-1}) light injuries are probable as a result of the undercarriage collapse. With a vertical component of up to 1,500 feet per minute (7.64 ms^{-1}) injuries are likely to be more severe, with some fatalities. Over 2,000 feet per minute (10.1 ms^{-1}) the likelihood of survival is small (see FAA AC 21-22 for a full description of the likelihood of injury as a result of impact).

To avoid exciting the phugoid mode, whenever possible pilots were advised not to attempt turning and pitching manoeuvres simultaneously.

Display evaluation trials

Participants

Ten line pilots, all holding a current ATPL (Airline Transport Pilot's Licence) and a type rating on the Boeing 747 were recruited to take part in the study. Seven were Captains and the remaining three were First Officers. The mean number of flying hours on all types was 13,610 hours (sd=5,776). The mean flying experience on all Boeing 747 variants was 6,298 hours (sd=3,824).

Experimental design

Participants were randomly assigned to one of two equally-sized groups. Group One first attempted four pre-defined tasks using only the aircraft's throttles as a means of control and without the benefit of the emergency flight control display system. They then repeated these manoeuvres with the benefit of the new display system. Group Two first completed the tasks with the new display and the repeated them without it.

The three evaluation tasks reflected the three modes available in the controlled flight by throttles emergency flight control display system. They were a recovery task; an en-route navigation task (which was comprised of two sub-tasks, a descent task and a turning task) and an approach and landing task. The tasks were always completed in this sequence, as this would be the order in which a pilot faced with a FCS malfunction would be required to undertake them. In these circumstances a degree of learning is actually desirable across the tasks, as the approach and landing task is the final (and most crucial) problem facing the pilot.

Evaluation tasks

Recovery task When this task was being undertaken under the guidance of the new display system, participants used the recovery mode. The participants were required to recover the aircraft to straight and level flight using only the throttles as a means of control. The task commenced at 20,000 feet (6,115m) when the aircraft was passing through a heading of 090° while holding a 20° angle of bank in a right hand turn. At the time of the simulated FCS failure the aircraft's trim speed was approximately 210 knots TAS (108.4ms-1). The task was deemed to be complete when the aircraft was within ± 5° of any stable heading and height oscillations were contained within ± 100 feet (30.6m). No final altitude or heading was specified. The measures taken were the time to regain control and the absolute deviation in the final altitude from the cruising altitude before the FCS failure.

En-route mode   The en-route flight task had two sub-tasks, a descent sub-task and a turning sub-task. The en-route mode was used when evaluating the emergency flight control display system.

For the descent sub-task the pilots were required to descend on a constant heading of 090° (inbound to a VOR). The task commenced at 11,500 feet (3,516m) AMSL (above mean sea level), 19.7nm DME (36.6km) from the beacon. The object was to descend to 8,500 feet (2,599m) on the 090° radial. To avoid exciting the phugoid pilots were advised not to use an FPA of greater than -2° (or approximately 750 feet per minute/3.8ms^{-1} vertical speed). The descent task was considered complete when the aircraft was within ± one dot (2°) on the course deviation indicator (CDI) of the 090° radial (either inbound or outbound) and was stabilised at the target altitude (±100 feet/30.6m) with oscillations contained within ±100 feet (30.6m). The measures taken were the lateral root mean square error (RMSE) from the 090° radial and longitudinal RMSE from the optimum 2° vertical profile descent path.

The descent task was followed by a turning task. This required a turn over a VOR beacon from the 090° (inbound) to the 120° radial (outbound) while maintaining a constant altitude of 10,000 feet (3,058). The task commenced at 5.0nm (9.3km) DME inbound to the beacon was considered complete when the aircraft was established within ±1 dot (2°) on the CDI on the 120° radial outbound and was also within ±100 feet of the assigned altitude with oscillations contained within ±100 feet. Lateral RMSE was calculated for deviations around the 120° outbound radial (starting at a distance of 2.5 miles/4.7km DME from the beacon) and vertical RMSE was calculated from the assigned altitude.

Approach and landing task   This commenced with the aircraft at 3,790 feet/1,159m AGL (4,000 feet/1,223m AMSL) on a heading of 082°, lined up with the runway's centreline at a distance of 25nm/46.5km. The task was complete when the aircraft touched the ground. The performance data logged were the RMSE from the localiser; the RMSE from the optimum 2° approach path; rate of descent at touchdown, and the distance of the touchdown point from the aiming point on the runway. The approach display mode was used in these trials.

On concluding each of the above tasks participants completed the rating component of the NASA-TLX workload scale (Hart and Staveland, 1988). The workload weighting component of the TLX was completed after each block of trials undertaken either with or without the benefit of the emergency display system. On finishing each trial the participant was also given a structured de-briefing.

Procedure

Prior to undertaking the trials all pilots were advised of their rights as a participant and gave their written consent. They were given a briefing about the nature of the research and the operation and logic of the emergency flight control display system. The briefing on the use of the system was not intended to be an extensive briefing. The level of information reflected the degree of familiarisation that a pilot may receive prior to a training session in a simulator or by reading the flight manuals. No opportunity was given to the participants to familiarise themselves with the system as no opportunity to do so would be available in a real emergency.

Following the briefing the participants were taken to the Cranfield University Engineering Flight Simulator to undertake the evaluation trials. The flight deck was configured in a similar manner to a Boeing 747-400 but had a Boeing 747-200 flight dynamic model. All pilots completed the three tasks described previously both with and without the benefit of the emergency flight control display system.

On completion of each trial the participant was given a thorough de-briefing about their performance and their opinions about the design and operation of the new system was solicited. Finally, the pilots were thanked for their time and paid £45 for participating.

Results

Recovery display mode

Recovering the aircraft to a straight and level attitude was performed significantly more quickly when using the recovery mode of the emergency flight control display system than when attempting the same manoeuvre without it ($F=29.31$; $df=1,8$; $p<0.001$): see table 1. There was no significant effect of performing the task with the new display system as either the first or second trial ($F=3.12$; $df=1,8$; $p>0.05$). There was also no significant interaction observed between use of the display system and trial order ($F=1.57$; $df=1,8$; $p>0.05$). In terms of *absolute* height deviations from cruising altitude during the recovery task there was no significant difference with respect to use (or not) of the emergency display ($F=2.16$; $df=1,8$; $p>0.05$), trial order ($F=0.77$; $df=1,8$; $p>0.05$) nor was there a significant interaction between these terms ($F=0.10$; $df=1,8$; $p>0.05$). There was, however, a significant reduction in overall workload experienced by pilots when recovering the aircraft to straight and level flight attributable to using the new display ($F=19.21$; $df=1,8$; $p<0.05$). However, there was no significant effect on

184 *Jason M. Demagalski, Don Harris and James E. Gantry*

workload attributable to trial order (F=1.34; df=1,8; p>0.05) nor was there a significant interaction term (F=0.29; df=1,8; p>0.05).

Table 1   **Measures to assess performance when recovering the aircraft to straight and level flight with and without the emergency display system. The display was used in either the first or the second set of trials. Time to recover is measured in seconds; mean absolute deviations from assigned cruising altitude are measured in feet.**

Display Condition	Trial Order	Mean time to recover (s.d.)	Mean absolute deviations (s.d.)	Overall NASA TLX score (s.d.)
Using emergency display system	*1st*	102.6 (14.9)	879.0 (203.3)	42.7 (16.2)
	2nd	77.2 (8.2)	542.6 (160.8)	52.8 (11.0)
	Overall	89.8 (17.5)	710.8 (247.2)	47.8 (14.1)
Without emergency display system	*1st*	326.4 (132.9)	1796.1 (1842.8)	70.3 (8.6)
	2nd	217.0 (88.9)	1139.9 (1551.2)	74.4 (15.2)
	Overall	271.7 (121.2)	1468.0 (1642.7)	72.4 (11.8)

En-route display mode

Descent task   There was a significant reduction in RMSE from the optimum vertical profile when using the emergency display system (F=9.51; df=1,8; p<0.05). Trial order (whether the new display system was presented in the first or second set of trials) had no significant effect (F=1.84; df=1,8; p>0.05) nor was there a significant interaction between use of the new display system to control the aircraft and trial order (F=0.46; df=1,8; p>0.05): see table 2. There was no significant difference observed in RMSE from the optimum horizontal track (F=1.41; df=1,8; p>0.05); no effect of trial order (F=0.01; df=1,8; p>0.05) nor was there a significant interaction term (F=4.83; df=1,8; p>0.05). Although the new display system encourages superior vertical track-keeping it did not help to reduce significantly the workload experienced by the pilots (F=3.44; df=1,8; p>0.05). As before, trial order also had no effect on workload (F=0.30; df=1,8; p>0.05) nor was there a significant interaction observed between order of presentation and the order in which the trials were conducted (F=0.73; df=1,8; p>0.05).

Table 2   **Measures to assess the performance during the descent sub-task when evaluating the en route mode of the emergency display system. The display was used in either the first or the second set of trials. Vertical and lateral root mean square error (RMSE) from optimum is measured in feet.**

Display Condition	Trial Order	Mean vertical RMSE (s.d.)	Mean lateral RMSE (s.d.)	Overall NASA TLX score (s.d.)
Using emergency display system	*1st*	412.2 (161.1)	8283.6 (2538.5)	55.2 (27.4)
	2nd	283.5 (172.7)	4596.9 (1736.3)	66.1 (10.0)
	Overall	347.9 (171.4)	6645.1 (2851.0)	60.6 (20.3)
Without emergency display system	*1st*	1292.2 (726.7)	6617.1 (3540.4)	78.3 (11.3)
	2nd	845.5 (642.0)	10185.5 (4203.7)	74.7 (14.3)
	Overall	1068.8 (688.0)	8203.0 (4053.5)	76.5 (12.3)

Turning task   There was no significant difference in vertical RMSE from the assigned cruising altitude when performing the turning task either with or without the emergency display system ($F=4.65$; $df=1,8$; $p<0.10$), although as will be noted from the manner in which the results are quoted, the result did border on significant suggesting that the display system did produce slightly superior performance (see table 3). Trial order ($F=0.08$; $df=1,8$; $p>0.05$) and the interaction term ($F=0.49$; $df=1,8$; $p>0.05$) were also both non-significant. As for the descent task, there were no significant differences in lateral track-keeping RMSE on the outbound (post-turn) radial attributable to the display condition ($F=0.01$ $df=1,8$; $p>0.05$) or the trial order ($F=1.29$; $df=1,8$; $p>0.05$). The interaction term was also non-significant ($F=1.76$; $df=1,8$; $p>0.05$). The display system did, however, significantly reduce the level of pilot workload during the task ($F=8.85$; $df=1,8$; $p<0.05$). As before, there was no effect observed attributable to trial order ($F=1.27$; $df=1,8$; $p>0.05$) nor was there a significant interaction term ($F=0.02$; $df=1,8$; $p>0.05$).

Approach and landing display mode

There was a significant improvement in vertical RMSE when using the emergency display system. Deviations from the optimum 2° approach profile were

186 *Jason M. Demagalski, Don Harris and James E. Gantry*

significantly smaller when using the new display than when pilots did not have the benefit of using it (F=12.67; df=1,8; p<0.01). There was also a significant effect of trial order. Participants performed significantly better on their second attempt irrespective of whether they had previously used the new display or not (F=8.09; df=1,8; p<0.05). The interaction term was bordering on significance (F=4.31; df=1,8; p<0.10) suggesting that there was a larger improvement in performance during the second trial in those pilots who attempted the approach and landing task without the benefit of the new display system in their first trial (see table 4). The effect of using the emergency display system in the lateral control axis was non-significant. There was no effect on performance attributable to using the display system (F=3.50; df=1,8; p>0.05) nor was there an effect of trial order (F=0.08; df=1,8; p>0.05). The interaction term between was also non-significant (F=0.58; df=1,8; p>0.05).

Table 3   **Measures to assess performance during the turning sub-task when evaluating the en route mode of the emergency display system. The display was used in either the first or the second set of trials. Vertical and lateral root mean square error (RMSE) from optimum outbound VOR track is measured in feet.**

Display Condition	Trial Order	Mean vertical RMSE (s.d.)	Mean lateral RMSE (s.d.)	Overall NASA TLX score (s.d.)
Using emergency display system	*1st*	225.6 (70.8)	6644.9 (4693.0)	40.9 (22.7)
	2nd	155.2 (49.8)	7722.6 (8318.7)	51.4 (14.8)
	Overall	190.5 (68.7)	7183.7 (6392.7)	46.2 (18.9)
Without emergency display system	*1st*	475.0 (255.6)	10467.2 (8269.1)	59.1 (12.4)
	2nd	644.1 (721.3)	3460.0 (1891.8)	67.9 (13.6)
	Overall	559.5 (517.9)	6963.6 (6754.3)	63.5 (13.1)

Pilots also landed significantly closer to the aiming point of the runway when using the new display system (F=6.57; df=1,8; p<0.05). Trial order, however, had no significant effect on this measure (F=0.29; df=1,8; p.>.05). The interaction term between display condition and trial order was also non-significant (F=0.88; df=1,8; p>0.05). These data can also be found in table 4. As in all other evaluations, the emergency display significantly reduced the overall workload experienced by pilots (F=8.62; df=1,8; p<0.05). As before, both trial order (F=0.19; df=1,8; p>0.05) and the interaction term (F=0.18; df=1,8; p>0.05) were non-significant.

Table 4   Measures to assess the performance of pilots when using the approach and landing mode of the emergency display system. The display was used in either the first or the second set of trials. Vertical and lateral root mean square error (RMSE) from the optimum approach path is measured in feet; distance from the aiming point of the runway is measured in nautical miles.

Display Condition	Trial Order	Mean vertical RMSE (s.d.)	Mean lateral RMSE (s.d.)	Distance from aiming point (s.d.)	Overall NASA TLX score (s.d.)
Using emergency display system	*1st*	361.2 (175.6)	1359.6 (924.0)	0.81 (0.52)	58.11 (28.4)
	2nd	402.7 (103.8)	2067.8 (624.0)	1.00 (0.39)	64.1 (10.6)
	Overall	784.3 (331.1)	1713.9 (831.7)	0.91 (0.45)	61.1 (20.5)
Without emergency display system	*1st*	998.1 (292.5)	3067.6 (2140.3)	3.52 (3.21)	81.5 (10.5)
	2nd	570.4 (216.2)	2789.2 (2012.3)	2.27 (2.14)	81.6 (14.7)
	Overall	784.3 (331.1)	2928.4 (1964.0)	2.90 (2.65)	81.6 (12.1)

From figure 3 it can be seen that when using the emergency flight control display system only one 'landing' exceeded a 2,000 feet per minute rate of descent. All other landings had a rate of descent of less than 1,500 feet per minute at touchdown and half the landings had a rate of descent of less than 800 feet per minute, suggesting that in these cases the undercarriage would have been unlikely to collapse. Without the display system half of the landings had a rate of descent in excess of 1,500 feet per minute, making death or serious injury a likely outcome for all on board.

Discussion

Results from the initial trials undertaken to evaluate the emergency flight control display system for flying an aircraft using only its throttles are extremely

188 *Jason M. Demagalski, Don Harris and James E. Gantry*

encouraging. They strongly suggest that pilots using the new system would be more likely to maintain control of a severely disabled aircraft. The system would also give them a reasonable chance of executing a technically-survivable landing within the environs of an airfield that may result in only relatively minor injuries to the passengers and crew. It does need to be stated, though, that these results are preliminary and more development and testing is required. However, the basic display concept and the operating logic would appear to be sound.

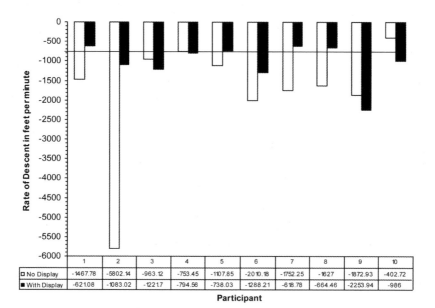

	1	2	3	4	5	6	7	8	9	10
□ No Display	-1467.78	-5802.14	-963.12	-753.45	-1107.85	-2010.18	-1752.25	-1627	-1872.93	-402.72
■ With Display	-621.08	-1083.02	-1221.7	-794.56	-738.03	-1288.21	-618.78	-664.46	-2253.94	-986

Participant

Figure 3   Rate of descent at touchdown (feet per minute) for each participant. Horizontal line is at 800 feet per minute.

There is a noticeable pattern across all the results. The major performance benefit in the control of a disabled aircraft using engine thrust alone accrues in the vertical (pitch) axis. In many cases, the results in the horizontal (lateral) axis show that the new display system offers only a marginal advantage (if any) over no display. However, given that the major problem in controlling an aircraft using thrust alone is in avoiding exciting the phugoid mode, it is not too surprising that it is in this area that the new system's greatest benefits lie. The reduction in RMSE when using the new display indicates that the new system is of benefit in either avoiding exciting the phugoid mode or in reducing the size of the

oscillations. Either of these effects would explain the greater precision in flying the aircraft by throttles alone when using the new display. Reducing the size of the oscillations is extremely desirable. It should be recalled that these oscillations are of extremely low frequency (in the order of one cycle per minute). It will not be possible to control precisely at what point in the cycle that the pilot may need to effect a landing hence s/he may need to land during the downward component. If the oscillations are particularly violent, the vertical speed on the downward component may be so great as to destroy the aircraft and seriously injure (or kill) its occupants, as at Sioux City. The results reported herein suggest that when using the emergency display system the vast majority of landings made were technically survivable. Even if the aircraft touched down on the downward phase of the phugoid, these oscillations were so reduced in amplitude when using the display that the landings were still survivable.

In addition to predicting the oscillatory motion of the phugoid, the lags when attempting to control an aircraft using engine thrust alone are the second major problem faced by a pilot. As the new display system had a predictor element incorporated into it in terms of the commanded throttle position required to achieve a certain altitude or heading, this overcame many of the control lag problems that induce poor performance. The display of predicted flight path, even if it were easily calculable and predictable, would be of little use as the pilot would not be able to exert such fine control using only the throttles as to be able to fly the desired flight path. The strategy of using only relatively gentle rates of climb and descent to avoid exciting these motions would seem to be vindicated, although this did have the effect of considerably further compromising the manoeuvrability of the aircraft (but at the benefit of maintaining stability). This limited manoeuvrability further added to the requirement for the incorporation of a predictive display element of flight path in addition to the command display element for throttle position, in order to fly with any reasonable degree of accuracy.

Controlling the aircraft is only one of the problems, though, faced by pilots in the event of an FCS failure. They will also need to select a suitable diversionary airfield, navigate the flight to their new destination and communicate with air traffic control and the cabin crew, amongst a myriad of other safety critical tasks. The results demonstrate that the emergency flight control display system considerably reduces the overall workload (as measured by the NASA-TLX) when attempting to fly the aircraft by the use of throttles alone. This implies that the crew will be more likely to have the spare cognitive capacity required to complete their other tasks more successfully.

Conclusions and future developments

Despite the limitations of the prototype interface, the flight by throttles emergency display system demonstrates that with a little assistance to the pilots, in the event of an FCS failure it is possible to maintain control of a large aircraft and make a survivable emergency landing using engine thrust alone as a means of control. The display system is still in its early stages of development. The initial design concept

190 *Jason M. Demagalski, Don Harris and James E. Gantry*

described in this paper was implemented on a four-engined heavy commercial aircraft, however the vast majority of the commercial fleet, including the next generation of medium/large commercial aircraft, consists of twin-engined aircraft. As a result the system needs to be developed to accommodate aircraft of this type. The need to develop it for three-engined aircraft, such as the McDonnell-Douglas (Boeing) DC10/MD11 series of aircraft, as used by NASA in its research, is far less pressing as this engine configuration is no longer used in jet transport aircraft currently in production. It should also be noted that this system's utility is restricted to 'conventional' configuration aircraft that exhibit natural aerodynamic stability. The system will not work on aircraft such as supersonic transport aircraft or future blended wing body configurations, neither of which is aerodynamically stable. Even 'conventional' configuration aircraft with relaxed aerodynamic stability (e.g. those with reduced tail area to minimise aerodynamic drag) cannot benefit from this system.

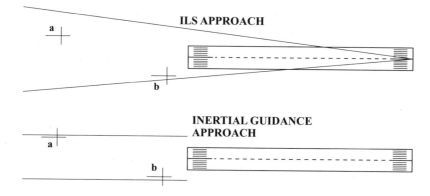

Figure 4   Lateral ILS approach versus lateral inertial guidance approach

It was also noted that there was a slight tendency to 'over-control' the aircraft on the later stages of approach. Consideration is being given to employing a display format based upon an inertial guidance display rather than a conventional localiser-type of display. The rationale for this is best described with reference to figure 4. In this figure aircraft A and B are both the same distance from the runway centreline, however aircraft B is at a much greater angular deviation from the aiming point than is aircraft A. As a result, an ILS-type display would show the approaching aircraft to be dramatically off-course. An ILS display has an increasing level of sensitivity for off-course deviations as the aircraft gets closer to the runway threshold however the fine control to respond to such commands is not available when attempting to control an aircraft using the throttles alone. The error displayed on a display based around an inertial guidance approach, however, relates simply to the distance from the runway centreline, perhaps discouraging the pilot from attempting to 'over-control' the aircraft during the final stages, hence reducing the likelihood of the aircraft becoming unstable.

At the moment, the command display of throttle position is presented head-down on the primary engine systems' display. Clearly this is not the optimum position for information relating to a manual control task of any kind, let alone one relating to flight path guidance. The increasing use of head-up displays in commercial aircraft, though, may provide the perfect location for the display of this information. This will also allow flight path vector information to be displayed to the pilot overlaid on the outside world. This should be particularly beneficial during approach and landing. These refinements will be evaluated in the near future.

References

Andre, A. D. and Wickens, C. D. (1990). *Display-Control compatibility in the cockpit: Guidelines for display layout analysis.* (Technical Report ARL-90- 12/ NASA A3I-90-1). University of Illinois Aviation Research Laboratory: Savoy, IL.

Blezad, D.J. (1996). The propulsive-only flight control problem. *NASA-CR-202408.* Ames Research Center; Author.

Bull, J., Mah, R., Hardy, G., Sullivan, B., Jones, J., Williams, D., Soukup, P. and Winters, J. (1997). Piloted Simulation Tests of Propulsion Control as Backup to Loss of Primary Flight Controls for a B747-400 Jet Transport. *NASA-TM-1997-112191.* Dryden: NASA.

Burcham, F.W. Jr., Fullerton, C., Gilyard, G., Wolf, T. and Stewart, J. (1991). A preliminary investigation of the use of throttles for emergency flight control. In, *Proceedings of the 27th AIAA/SAE/ASME/ASEE Joint Propulsion Conference, Sacremento CA June 24-26.* AIAA paper 91-2222. Washington D.C.: American Institute of Aeronautics and Astronautics.

Burcham, F.W., Maine, T.A. and Burken, J.J. (1998). Using Engine Thrust for Emergency Flight Control: MD-11 and B-747 Results. *NASA-TM-1998-206552.* Dryden: NASA.

Burcham, F.W., Maine, T.A., Kaneshige, J. and Bull, J. (1999). Simulator Evaluation of a Simplified Propulsion-Only Emergency Flight Control Systems on Transport Aircraft. *NASA-TM-1999-206578.* Dryden: NASA.

Federal Aviation Administration (1985). *Injury Criteria for Human Exposure to Impact (AC 21-22).* Washington DC: US Department of Transportation.

Hart, S.G. and Staveland, L.E. (1988). Development of a Multi-dimensional Workload Rating Scale: Results of Empirical and Theoretical Research. In, P.A. Hancock and N. Meshkati (Eds) *Human Mental Workload.* Amsterdam: Elsevier.

Maine, T., Schaefer, P., Burken, J. and Burcham, F. (1994). Design Chalenges Encountered in a Propulsion-Controlled Aircraft Flight Test Program. In, 30th *AIAA/ASME/SAE/ASEE Joint Propulsion Conference (AIAA paper 94-3359).* Indianapolis, IA, June 27-29.

McRuer, D.T., Hoffman, L.G., Jex, H.R., Moore, G.P., Phatak, A.V., Weir, D.H. and Wolkovitch, J. (1968). *New Approaches to Human-Pilot/Vehicle Dynamic Analysis* (AFFDL-TR-67-150). Dayton, OH: Wright Patterson AFB.

192 *Jason M. Demagalski, Don Harris and James E. Gantry*

National Transportation Safety Board (1990). *Aircraft Accident Report: United Airlines Flight 232, McDonnell Douglas DC-10-10. Sioux Gateway Airport, Sioux City, Iowa, July 19, 1989. Report PB90-910406, NTSB/AAR 90/06 (November).* Washington DC: Author

Wickens, C.D. (1992). *Engineering Psychology and Human Performance (2nd Edition).* New York: HarperCollins.

Acknowledgements

The contribution made by Roger Bailey, the Chief Test Pilot in the College of Aeronautics, Cranfield University, in the design and development of the emergency flight control display system, is gratefully acknowledged.

Index